Josiah Stickney Lombard

Experimental Researches on the Regional Temperature of the Head

Under Conditions of Rest, Intellectual Activity and Emotion

Josiah Stickney Lombard

Experimental Researches on the Regional Temperature of the Head
Under Conditions of Rest, Intellectual Activity and Emotion

ISBN/EAN: 9783337373191

Printed in Europe, USA, Canada, Australia, Japan

Cover: Foto ©berggeist007 / pixelio.de

More available books at **www.hansebooks.com**

EXPERIMENTAL RESEARCHES

ON THE

REGIONAL TEMPERATURE

OF

THE HEAD

UNDER CONDITIONS OF REST, INTELLECTUAL ACTIVITY, AND EMOTION.

BY

J. S. LOMBARD, M.D.,
FORMERLY ASSISTANT PROFESSOR OF PHYSIOLOGY IN HARVARD UNIVERSITY, U.S.

LONDON:
H. K. LEWIS, 136, GOWER STREET.
1879.

PREFACE.

THE experiments contained in this book were commenced early in January, 1877, and were pursued with but slight interruptions until March of the present year, the total number of observations made in this time being upward of sixty thousand. These observations have been carefully sifted, and only those accepted which were made under conditions where it was certain that the greatest caution had been exercised to avoid all sources of error.

Nearly the whole of Parts I and II, and a small portion of Part III, were communicated to the Royal Society in March and June, 1878.* Since the time of these communications, the author has made a great number of additional observations, and has found reason to modify slightly some of the views then expressed on the effect of mental activity on the temperature of the head. The unavoidably sudden and premature closing of the investigations has left unsolved two questions which the author had intended specially to examine, namely, the relative temperatures, in the quiescent mental condition, of different portions of the surface of the head, taken in larger areas than the subdivisions adopted in the present experiments; and the effect of *simple attention* to visual and auditory impressions on the temperature of different parts of the head; this latter being an enquiry for the

* 'Proceedings of the Royal Society,' No. 186, p. 1, March 7th, 1878; No. 188, p. 457, June 20th, 1878.

suggestion of which the author is indebted to Dr. Charlton Bastian.

The author's thanks are due to Dr. Brown-Séquard and to Dr. Bastian for valuable advice. He is also greatly indebted to Dr. Frederic H. Haynes, of Leamington, for much personal assistance, many of the most important results obtained being, in a great measure, due to the unfailing patience and intelligent interest shown by Dr. Haynes.

LONDON; *December*, 1879.

CONTENTS.

	PAGE
PREFACE	iii
INTRODUCTION	1

PART I.

DESCRIPTION OF APPARATUS EMPLOYED AND EXPERIMENTAL METHODS ADOPTED IN THE EXAMINATION OF THE TEMPERATURE OF THE HEAD.

CHAPTER I.

Thermometers.—Currents in thermo-piles.—Phenomena of reverse currents.—Inequality of piles,—its causes, detection, and remedy . 4

CHAPTER II.

Methods of testing the relative temperatures of two parts.—Different movements of the galvanometer needles, and their causes.—Composition and construction of thermo-piles employed.—Rheostat and keys.—Galvanometers 16

PART II.

THE RELATIVE TEMPERATURES OF DIFFERENT PARTS OF THE SURFACE OF THE HEAD IN THE QUIESCENT MENTAL STATE.

CHAPTER I.

General remarks.—Divisions and measurements of the head.—Subdivisions and measurements of the anterior region.—Examination of the anterior region in symmetrically situated spaces of the two sides . 27

CHAPTER II.

Examination of the anterior region in spaces on one and the same side . 45

CHAPTER III.

Subdivisions and measurements of the middle region.—Examination of the middle region in symmetrically situated spaces of the two sides . 56

CHAPTER IV.

Examination of the middle region in spaces on one and the same side . 68

CHAPTER V.

Subdivisions and measurements of the posterior region.—Examination of the posterior region in symmetrically situated spaces of the two sides 79

CHAPTER VI.

Examination of the posterior region in spaces on one and the same side. —Comparison of the three regions with each other.—Absolute temperatures of the three regions 89

CHAPTER VII.

Causes of disturbance affecting the experiments.—Considerations regarding the influence of the temperature of the brain on the temperature of the outer surface of the head.—Propagation of slight differences of temperature in bad conductors.—Effect of the circulation of the blood on the outward transmission of heat from the brain to the integument.—General conclusions regarding a correspondence between the relative temperatures of points of the integument and the relative temperatures of underlying points of the brain surface . 107

PART III.

THE EFFECT OF INTELLECTUAL AND EMOTIONAL ACTIVITY ON THE TEMPERATURE OF THE HEAD.

CHAPTER I.

General considerations.—Effect of different kinds of intellectual work on the temperatures of the three regions 120

Contents. vii

CHAPTER II.

PAGE

Experiments illustrative of the effect of intellectual work on the temperature of different parts of the head.—Average rises of temperature for the different regions, in intellectual work.—Rises of temperature above the average, in intellectual work 134

CHAPTER III.

Comparative effect of intellectual work on the temperature of different spaces of one and the same side of the head . . . 146

CHAPTER IV.

Reversals of the usual order of superiority of rise of temperature in different spaces in intellectual work.—Effect of verbal expression in intellectual work on the comparative rise of temperature in the surface over Broca's convolution.—Precautions respecting the absolute temperatures of the parts examined 159

CHAPTER V.

Comparative effect of intellectual work on the temperature of symmetrically situated spaces of the two sides of the head . . 168

CHAPTER VI.

General effect of emotional activity on the temperature of the three regions.—Experiments illustrative of the effect of emotional activity on the temperature of different parts of the head.—Average rises of temperature for the different regions, in emotional activity . . 175

CHAPTER VII.

Comparative effect of emotional activity on the temperature of different spaces of one and the same side of the head.—Comparative effects of emotional activity with and without verbal expression . . 183

CHAPTER VIII.

Comparative effect of emotional activity on the temperature of symmetrically situated spaces of the two sides of the head.—Examples of the effects of anger, vexation, and mirth, on the temperature of the head 196

ERRATA.

Page 65, in 4th tier, *for* (0·594° F.), *read* (0·0594° F.).
,, 79, line 2, *for* page 37, *read* page 29.
,, 84, ,, 1, ,, percentages of, *read* percentages for.
,, 99, in heading, average percentage of times of occurrence of superiority of temperature of either side of the head, *add* over the other.
,, 103, in 1st district, 4th tier, right side, *for* 92·743° F., *read* 92·746° F.
,, 105, in 3rd district, 1st tier, left side, *for* 92·693° F., *read* 92·663° F.
,, 106, line 10, *for* (92·267° F.), *read* (92·276° F.).
,, 107, ,, 28, ,, p. 35, *read* p. 28.
,, 125, ,, 32, ,, (55·98° F.), *read* (54·98° F.).
,, 173, ,, .10, ,, (0·00892° C.), *read* (0·00892° F.).
Page 22, line 17, *for* (4·43) *read* (0·433).
,, 91, ,, 18, under left side, *for* 60·00 *read* 0.
,, 93, ,, 17, *put* 6th tier *after* 3rd tier.
,, 105, in 3rd district, 6th tier, right side, *for* 92·93°F. *read* 91·553° F.
,, 106, line 11, *for* 33·785°C. *read* 33·385°C.
,, 109, ,, 18, *after* region *put a period*.
,, 111, ,, 38, *for* too *read* two.
,, 111, ,, 27, *after* heat *put a period*.
,, 200, ,, 20, *for* (0·0116°F.) *read* (0·0101°F.).
Appendix, line 3, *omit* figure 3.

THE
REGIONAL TEMPERATURE OF THE HEAD.

INTRODUCTION.

THE investigations contained in this work had in view the following primary objects:—

1st. To determine, as far as possible, by a great number of carefully made observations, the normal relative temperatures of different portions of the surface of the head, when the brain is comparatively inactive.

2nd. To study the effect of different mental states on the temperature of different portions of the surface of the head.

The ultimate objects were also two,—namely:—

1st. To furnish, if possible, some reliable data as a starting point for examining the temperature of the surface of the head in morbid cerebral conditions.

2nd. To see if, from an examination of the relative temperatures of different portions of the surface of the head during increased mental action, any information could be obtained as to the comparative importance of the parts played by different portions of the surface of the brain in intellectual work and in the various emotions.

With regard to the first of these ultimate objects, it is easy to see that everything depends upon an accurate knowledge of the normal relative temperatures of the different portions of the surface of the head, and the variations of such temperatures within the limits of health. Without this preliminary knowledge, the examination of the temperature of the surface of the head can lead to no definite conclusions as to the existence of localized cerebral disorder.

So far as the author is aware, Dr. Wm. A. Hammond of New York, was the first to indicate that a difference of temperature between the two sides of the head exists in health. In 1875, Dr. Hammond, employing a thermo-electric apparatus devised by the author the year previous, came to the conclusion, from observations made on a large number of individuals, that the left side of the head has a higher temperature than the right side.* Unfortunately, the abstract of Dr. Hammond's paper, in the author's possession, does not specify the *exact part* of the head experimented on,—a matter of great importance, as will be seen farther on. It is, therefore, impossible to compare Dr. Hammond's results with those to be given in the present work.

In 1877, M. Broca communicated to the French Medical Association the results of experiments made by him, with thermometers, on the relative and absolute temperatures of different parts of the surface of the head. M. Broca states that in the normal quiescent mental and physical condition, the temperature is always higher on the left side than on the right side, in the frontal, temporal, and occipital regions. He has also found that the frontal region has normally the highest absolute temperature, the temporal region coming next, and the posterior region last, in point of thermal activity.†

It will be seen that the author's investigations lead to conclusions differing materially from those arrived at by M. Broca.‡

With regard to the second of the ultimate objects of the present investigations, namely, the relative effect on the temperature of different parts of the surface of the head of intellectual and emotional activity, it is evident that here also a thorough acquaintance with the relative temperatures of the different parts of the surface of the head in the quiescent mental condition is of value, although not absolutely essential, as a large part of the investigations on this subject may be pursued independently of any

---

* Dr. Hammond's paper was read at a meeting of the New York Neurological Society, October 4th, 1875.

† 'Revue Scientifique,' September 15th, 1877.

‡ Dr. L. C. Gray of Brooklyn, N. Y., has also investigated, still more recently, this subject. ('New York Medical Journal,' August, 1878.)

Introduction.

particular knowledge of the relative temperatures of the parts experimented on, when the brain is inactive.

What has been already done in the subject of the effect of mental activity on the temperature of the head may be summed up in a few words.

In 1866 the author commenced a series of experiments, with thermo-electric apparatus, on the temperature of the human head in the quiescent mental condition, and in the states of intellectual and emotional activity. These experiments showed that the exercise of the higher intellectual faculties, as well as the different emotions, caused a rise of temperature in the head perceptible through the medium of delicate apparatus. Merely arousing the attention could produce the same result. These investigations were published in June 1867.* Toward the close of the latter year, Professor Moritz Schiff, who had been working independently of any knowledge of what the author had been doing, communicated to the Museum of Natural History of Florence results of a similar nature.† In 1870 Professor Schiff published an account of a series of investigations made directly upon the brain of animals, which decisively proved that mental exertion is accompanied by an elevation of temperature in the brain. M. Broca, in the investigations already alluded to, has also found an elevation of temperature, in the human head, during increased mental activity; and still more recently, M. Paul Bert has likewise arrived at a similar result.‡

In this work, after examining at some length the instruments employed and the experimental methods adopted in the investigations, we will consider, first, the normal relative temperatures of different parts of the surface of the head, taken in small spaces, in the inactive mental state; and, second, the changes of temperature in these spaces produced by intellectual and emotional activity.

---

* 'New York Medical Journal,' June, 1867; and 'Archives de Physiologie,' September—October, 1868.

† 'Archives de Physiologie,' t. iii, p. 6, 1870.

‡ 'British Medical Journal,' April 19th, 1879.

PART I.

DESCRIPTION OF APPARATUS EMPLOYED AND EXPERIMENTAL METHODS ADOPTED IN THE EXAMINATION OF THE TEMPERATURE OF THE HEAD.

CHAPTER I.

THERMOMETERS.—CURRENTS IN THERMO-PILES.—PHENOMENA OF REVERSE CURRENTS.—INEQUALITY OF PILES—ITS CAUSES, DETECTION, AND REMEDY.

BOTH thermometers and thermo-electric apparatus have been employed in these experiments, but thermo-electric apparatus has been principally used. The author does not consider thermometers very reliable in investigations of the kind in question, for the reason, that they cannot be pressed down firmly enough upon the surface to empty the superficial vessels. The best form of this instrument, for the purpose in hand, with which the author is acquainted, is the surface thermometer of Dr. E. Seguin of New York.* When ordinary thermometers are used, the side of the bulb not in contact with the skin is protected by a cap of cork lined with wool. In the experiments on the effect of mental activity, thermo-electric apparatus is, in the majority of cases, indispensable, to follow the slight and irregular changes in temperature which occur.

With regard to the thermo-electric apparatus made use of, much more must be said. The successful employment of delicate apparatus of this kind, in such investigations as we have to deal with, requires the greatest care and a good deal of practical experience. We will therefore consider somewhat fully the

* Dr. L. C. Gray has improved Dr. Seguin's thermometer. (Op. cit.)

ways of experimenting with thermo-electric apparatus, and the errors to be avoided in its use.

We will start with an examination of certain fundamental principles of thermo-electric piles, which have an important bearing on the use of the latter in such experiments as the present, and which are not treated of in ordinary works on physics.

THERMO-ELECTRIC PILES.

In a single thermo-electric pair composed of two metals, with conducting wires of the same metal each as the other, but different from those of the pile, we have three currents, namely,— the current proper of the pile, arising from the heating or cooling of the junction of the two metals forming the pile,—and a current produced by the heating or cooling of the junction of each of the two conducting wires with the metals of the piles,—these latter currents arising secondarily to the first current, by conduction from or to the junction proper of the pile.

Suppose we have a pile composed of bismuth and antimony, with copper conducting wires;—when heat is applied to the bismuth-antimony junction, a current is developed, passing across the junction from the bismuth to the antimony, the current thus produced being equal to 35, compared with a silver-copper pair as unity. This current will have a strength proportionate, within certain limits, to the difference of temperature between the pile, at the moment of the experiment, and the source of heat applied; and as the pile, in its quiescent state, will be at the temperature of the air, the strength of the current will be a measure of the difference of temperature between the air and the source of heat applied.

After a longer or shorter time—according to various circumstances to be hereafter examined—the conduction of heat from the junction of the pile to the farther ends of the bismuth-antimony bars, comes to affect the points of the latter with which are connected the copper conducting wires, and at each of these points a current is developed. In the bismuth-copper junction, the current goes from the bismuth to the copper through the junction, thus opposing the current proper of the pile; in the antimony-copper junction, the current goes from the copper to the antimony through the junction, thus also opposing the current proper of the pile. The bismuth-copper junction gives a current equal to 24 (using the same standard of comparison as above); and the antimony copper-junction gives a current equal to 11;—the two together, therefore, giving a total current equal to that of the bismuth-antimony pair. If, then, the two farther ends of the bars composing the pile were heated to the same degree as the bismuth-antimony junction, the opposing currents would neutralise each other, and no effect would be produced on a galvanometer introduced into the circuit of the pile and its wires. As it is, however,—unless the bars be extremely short,—the bismuth-antimony junction receives more heat per unit of time than is conveyed to the other two junctions by conduction, so that the current proper of the pile predominates. Of course if cold instead of heat be applied to the pile, the above-mentioned currents have their directions reversed, but the principle remains the same.

Suppose now a pile, such as has just been described, with a galvanometer in the circuit, to have its exposed face, $i.\,e.$ the bismuth-antimony junction, applied to the surface of the body. We will have, first, the effect on the galvanometer of the current proper of the pile, which, we will suppose, keeps the needle at 20° deflection when its oscillations have come to an end. The needle does not, however, long remain at this point; sooner or later, it begins to move backward toward zero; and finally takes up a position between the latter point and 20°.* This backward movement of the needle is due to two causes, one of which exists in the pile itself and the other in the tissues examined: at the present moment we are concerned only with the former cause. The backward movement of the needle, so far as the pile is concerned, is the result of the secondary currents excited in the bismuth-copper and antimony-copper junctions by the conduction of heat along the bars of the pile from its face. If the supply of heat to the face of the pile be small, but little will reach the remote junctions, and the retrogression of the needle will be long in showing itself, and will be of limited extent. If, however, the supply of heat be plentiful,—as is usually the case with warm-blooded animals,—an approach to equalisation of temperature in

* It frequently happens that the retrogression of the needle commences immediately on the cessation of its oscillations, there being no preliminary pause.

the whole pile will soon take place, and the retrograde movement of the needle will be prompt and decided.*

The rapidity and extent of retrogression of the needle is also influenced by the following conditions, namely :—

The length and thickness of the bars composing the pile.

The degree in which the bars are protected in their course, beyond the face of the pile, from the surrounding atmosphere and neighbouring bodies.

The conducting powers and specific heats of the metals composing the piles.

Thus long bars will, other things being equal, give a less prompt and less extended retrogression of the needle than short bars ; and the same is true of bars of small sectional area compared with thicker ones.†

Again, bars thinly covered, will give a less prompt and extended retrogression of the needle than those thickly enveloped in badly-conducting material ; for, in the latter case, the loss of heat to the surrounding medium being reduced to a minimum, there is a tendency to complete equalisation of temperature throughout the whole pile.

Further, bars composed of substances of high conducting power will give a prompter and more extended retrogression of the needle

* In piles composed of bad conductors of heat, and in which the bars are well protected on their sides and farther ends from the surrounding medium, the reverse or secondary currents may be made to manifest themselves very strikingly in the following manner :—The pile having been subjected for some little time to a source of heat considerably above the temperature of the air, is placed in a protected situation, and left to itself. The deflection of the needle diminishes gradually to zero,—when we see the needle, *instead of stopping at this point, pass steadily along on the opposite side of the scale to a considerable distance.* The reason of this peculiar movement, is, that the face of the pile, being more exposed to the air than the farther junctions, attains the temperature of the surrounding medium more quickly than do the remote ends of the bars,—hence the current proper ceases before the reverse currents have come to an end, and the latter carry the needle over to the opposite side of the scale.

† If the bars composing the pile be thoroughly protected from the surrounding atmosphere and neighbouring bodies, it will be found that the statement that thin bars show the reverse currents less quickly than thick bars is not correct. Where the bars are not protected from cooling influences, the bar of small sectional area loses heat more rapidly than the one of large sectional area ; but when well covered with non-conducting material the thinner bar will become heated throughout its whole length more quickly than the thicker bar by reason of its smaller mass.

than bars made of substances of low conductivity ; and, on the contrary, bars composed of substances possessing high specific heats will give, other things being equal, a more tardy and less extended retrogression of the needle than bars composed of substances having a low specific heat.

It will be seen from what has been stated, that we have our choice of two ways of observing differences of temperature with the galvanometer,—namely, either to take the deflection at the moment when the needle first comes to rest, before the secondary currents have time to develop themselves, or to wait until these currents have reached their maximum, and the whole pile has attained a permanent thermal condition. Of these two courses the latter is usually the better, but it is more tedious than the former, and consequently inconvenient when it is desirable to make a number of observations in a short space of time.

But there is a point now to be considered, bearing more or less upon all varieties of piles, which demands serious attention. It is this :—It is very difficult to make piles of equal current strengths, where great sensitiveness is required. The author has sometimes, *by accident*, obtained piles of equal powers, but no amount of care in the casting or putting together of the bars seems to ensure equality.

This inequality is due both to differences of resistance and differences of electro-motive force. The first-mentioned difficulty is, of course, got over readily by putting both piles in the same circuit; but the second difficulty is not so easily obviated. We must, in the first place, put the two piles in parallel circuits, and then introduce into the circuit of the one which has the stronger current a sufficient resistance to reduce the strength of the current to equality with that of the weaker pile.*

To do this, we expose the faces of the two piles to the same degree of heat or cold, and, waiting until the retrogression of the needle has come to an end, then introduce the requisite resistance in the circuit of the stronger pile to bring the needle to zero. The piles are now equal as regards their *permanent thermal condition;* but, unfortunately, it does not by any means follow that they will be found to possess equal current strengths when examined at a period *anterior* to the establishment of the permanent

* We are here supposing the use of two piles, for purposes of comparison, with the directions of their currents opposed to each other.

thermal state. If examined at the moment when the needle first comes to rest, and before the secondary currents manifest themselves, it will often be found that the resistance which produces equality when the permanent thermal condition is established, *is not the correct resistance for this earlier stage.* The reason of this is,—*that the electro-motive forces of the secondary currents do not, in the two piles, bear the same proportions to their respective currents proper.*

Thus, suppose the currents proper of the two piles to be represented by 10 and 5, respectively ;—a certain resistance introduced into the circuit of the stronger pile reduces its current strength to 5, —the same as the other. But now suppose that the reverse currents bear the proportions of 7 and 4; the resistance introduced has halved the strength of the secondary currents in the stronger pile, so that now these currents are represented by $3\cdot5$ instead of 7. When the secondary currents in both piles begin to act, the effective strength of the current proper of the first pile will be, $5-3\cdot5 = 1\cdot5$; while the effective strength of the current proper of the second pile will be, $5-4 = 1$; the two piles are consequently unequal at this phase of the observation. Of course, reversing the order of things, and making the piles equal after the establishment of the secondary currents, would be to render them unequal in the first phase of the observation.

Such a state of things as that just described frequently exists in thermo-piles. It is therefore necessary to vary the resistance of the circuit of the stronger pile according to the phase of the observation ; and the proper limits for the employment of each resistance must be determined by a careful testing, for each pair of piles.

We will now examine the methods adopted in the experiments made to detect and correct inequalities of current strength in the two piles.

To know if our piles are equal in strength, we must expose them both to the same constant source of heat or cold. This may be done in several ways. We may immerse the faces of the piles in water or oil having a temperature above that of the air ; or we may immerse them in melting ice. It is very difficult, however, to keep the temperature of the water or oil constant when such temperature differs considerably from that of the air, no matter what appliances are used ; and if rapid variations of temperature occur, the two piles may not act with equal quickness, even when their final current strengths are the same. With melting ice these difficulties are avoided.

Water simply exposed to the air in a room of pretty even temperature will keep at from half a degree to a degree centigrade below the air. By carefully watching the temperature and degree of moisture of the air of the room, we can keep the water at a sufficiently constant temperature for testing the equality of our piles. This would seem, therefore, a ready and safe method of testing;—but although correct enough under certain circumstances, it does not embrace all the conditions essential to an accurate determination of the respective strengths of the piles under circumstances where they are exposed to temperatures differing much more from the temperature of the air, than the difference between the latter and the water, in the above experiment. It must be understood that when two piles, opposed to each other, are exposed to a temperature differing from that of the air, each pile produces a current proportionate to the degree of difference existing between the temperature of the air and the new temperature affecting the face of the pile. If the piles possess equal power, exposure to the same degree of heat or cold will produce equal opposing currents, and the needle of the galvanometer will be unaffected. But if one pile is stronger than the other, the current of the former will predominate by an amount equal to the difference between the independent current-strengths of the two piles. Thus if one pile, acting alone, will, for a given thermometric difference, produce a deflection of the needle of $20°$ of the galvanometer scale, and the other pile, acting alone, will, for the same difference of temperature, produce a deflection of $10°$, the two piles, acting at the same time, in opposition to each other, will give a deflection of $10°$ in favour of the first pile.

Suppose now the temperature of the water in which the faces of the piles are immersed to be $9°$ C. ($48.2°$ F.), and that of the air to be $10°$ C. ($50°$ F.); and that, for a difference of $1°$ C. ($1.8°$ F.), one pile gives a deflection of $100°$ of the galvanometer scale and the other pile gives a deflection of $99.75°$. If we are using a Thomson galvanometer, $0.25°$ cannot be read on the scale, and consequently the needle would, to all appearances, be unaffected when the two piles were opposed to each other, and we would, therefore, conclude that they were of equal strength.

But now suppose the piles exposed to a temperature of $36°$ C.

(96·8° F.), the temperature of the air remaining the same as before. As the piles in the former instance gave, for a difference of 1° C., deflections of 100° and 99·75° of the galvanometer scale, respectively,—so now for a difference of 26° C. (46·8° F.), the first pile will give a deflection of 2600°, and the second pile a deflection of 2593·5°; and the two acting together against each other, will give a deflection of 6·5° in favour of the first pile.

This shows us the necessity, in testing the equality of piles, of employing temperatures differing as much from the temperature of the surrounding medium as the latter usually differs from the animal temperature.

The author has found the human body itself to furnish the most reliable means of testing the equality of piles.

We may test by means of the surface of the body in two ways. The first way is as follows:—

A space is selected on the skin,—usually on the thigh,—and surrounded with cotton wool and flannel, leaving only a small opening in the coverings just large enough to admit the faces of the two piles placed closely side by side. After the piles are applied, several minutes are allowed to elapse before the circuit is closed, in order to give time for the establishment of the permanent thermal condition in the piles.*

This apparently simple procedure, in reality requires much more care and time than would, at first sight, seem necessary. To begin with, *it is not always easy to find two points on the skin which have exactly the same temperature, when examined with delicate apparatus, even when these points are not more than a centimetre* (0·3937 inch) *apart*. Certain parts of the body are, however, of more uniform temperature than others, and experience has pointed out the anterior surface of the thigh as one of the regions best suited to the purpose in view.

The piles are previously tested in water to obtain an approximate estimate of the difference between them, and the experiment is then conducted as before described.

If after closing the circuit, we have a deflection, this deflection may be due either to a difference in the strength of the piles or to a difference of temperature in the two points on which the piles are placed, or to both these causes combined. To ascertain which of the causes is chiefly or solely instrumental in producing the

* The piles are applied with handles described farther on.

deflection, we reverse the positions of the two piles, placing each in the former position of the other. If the sole influencing cause of the deflection be a difference in the strength of the two piles, the needle will still be deflected, in the same direction, and to the same extent, as before; but if the deflection be due entirely to a difference in the temperature of the two points examined, the needle will now be deflected to the side of the scale opposite to that to which it was first deflected, the extent of the deflection being the same as in the first instance. It will frequently be found, however, that the deflection is due to both the causes specified combined. If now, the deflection be slight, we remove the coverings from the thigh and place over the region under examination, a piece of heavy felt, some 100 mm. (3·93 inches) square, which is fastened to the limb by means of tapes. In this felt there is an opening just large enough to admit the ends of the piles when the latter are placed as closely together as possible. On the under surface of the felt, around the opening, and extending for a centimetre (0·3937 inch) from its borders, is a thin copper plate, the object of which is to facilitate equalization of temperature in the two points upon which the piles are to be placed, by connecting them by a good conductor. The piles are now applied, and the limb carefully wrapped up. After some minutes, the connection is made with the galvanometer, and the deflection noted. The piles are now shifted, each over to the place of the other, and the deflection again noted. If the deflection persists in the same direction, and to the same extent, we know that the cause of the inequality lies in the piles themselves, and we introduce into the circuit of the stronger pile the requisite resistance to bring the needle back to zero. If, however, the deflection is not the same in direction and degree, and is, therefore, partly, at least, due to a difference of temperature between the two places examined, we must first determine, as closely as possible, how much of the deflection is due to each of the two causes enumerated.

Let us designate the two piles by "a" and "b," respectively, and the two points of the thigh by "A" and "B," respectively; and let "a," in the first instance, be placed upon "A," and "b" be placed upon "B." Further, let us suppose that, for the same difference of temperature, the strength of "a" is equal to 3, and that of "b" is equal to 2; and that "A" and "B" have the same temperature, which we will call 2. With "a" on "A," and "b" on "B," the independent currents of the two piles will be respectively, $3 \times 2 = 6$; and, $2 \times 2 = 4$; and with the two piles acting at once against each other, the deflection will be, $6-4 = 2$, in favour of "a." Changing the positions of the piles will make no difference, the balance in favour of "a" still remaining the same.

Next, let the piles be equal in strength, and the two points of the thigh of unequal temperature. We have here simply to transpose the values representing the strengths of the piles and the temperatures of the points of the skin, so that, "A" is equal to 3, "B" equal to 2, and "a" and "b" each equal to 2. In this arrangement the deflection will pass from one side to the other, according as "a" or "b" is, in turn, placed upon the hotter point "A." From these two cases we derive the following rules:

RULE 1.—When, on shifting the piles into each other's places, the direction and extent of the deflection remain the same, the cause of the inequality is the greater strength of the pile on the side of which the deflection exists.

RULE 2.—When, on shifting the piles into each other's places, the deflection passes over to the other side of the scale, its extent remaining the same, the inequality is due to a difference of temperature between the two points of the body, the hotter being the one on which that pile is placed which corresponds to the direction of the deflection at the given moment.

But now suppose "a" to be equal to 8; "b" to be equal to 4; "A" to be equal to 3, and "B" to be equal to 2. With "a" on "A," and "b" on "B," the independent currents will be, respectively, $8 \times 3 = 24$; and $4 \times 2 = 8$; and with the two currents opposed, there will be a deflection of $24-8 = 16$ in favour of "a." After shifting the piles, the independent currents will be, respectively, $8 \times 2 = 16$; and $4 \times 3 = 12$; and the deflection, when the two currents are opposed, will be 4, still in favour of "a." If we cause the values of "a" and "b" to change places with those of "A" and "B," respectively, we have with "a" on "A," and "b" on "B," the independent currents, "a" $= 3 \times 8 = 24$; and "b" $= 2 \times 4 = 8$, the same as before, and the same deflection of 16 in favour of "a," when the two currents are opposed to each other. But when the piles are shifted, the independent currents become, "a" $= 3 \times 4 = 12$, and "b" $= 2 \times 8 = 16$; hence, when the two currents are opposed, the needle passes over to the side of "b," and takes up a position of 4. From these cases we deduce two more rules, namely:

RULE 3.—If, on shifting the piles into each other's places, the deflection remains on the same side of the scale, although diminished or increased, the inequality is more owing to differences in the strength of the piles than to differences in the temperature of the parts examined.

RULE 4.—If, on shifting the piles into each other's places, the deflection passes over to the opposite side of the scale, its extent being either diminished or increased, the principal cause of the inequality is a difference of temperature between the two parts examined.

We have not unfrequently to do with still another condition of things, in which the action of the needle is different from anything thus far described. This is what occurs in the cases in question:—After applying the piles for the proper time, we find on closing the circuit that the needle remains motionless at zero: if, however, we reverse the positions of the piles we see the needle move up on one side or other of zero, and remain permanently deflected. This happens when the current strengths of the two piles, although different from each other, are yet in the same proportion to each other as are the temperatures of the two points examined to each other. Thus suppose "a" to be equal to 8, "b" to be equal to 4, "A" to be equal to 10, and "B" to be equal to 5 ; and that, in the first instance, " a " be placed on " B," and " b " be placed on " A ;" the current of " a " is now 8 × 5 = 40, and the current of " b " 4 × 10 = 40,—hence there is no deflection when the two currents are opposed to each other. On shifting " a " to " A," and " b " to " B," the current of " a " becomes 8 × 10 = 80, and the current of " b " becomes 4 × 5 = 20,—consequently when the two currents are opposed there is a deflection of 60 on the side of " a." From the preceding results, we obtain the following additional rule :

RULE 5. If the needle remains at zero at the first observation, and, on shifting the piles into each other's places, is deflected to one side of zero, the two piles differ from each other in strength in the same proportion as that in which the two parts examined differ from each other in temperature; and the pile corresponding to the side of the scale on which the deflection takes place, and the part to which this pile is applied, are, respectively, the one stronger in current, and the other higher in temperature, than the other pile and part.

If, after applying the above rules, we find that the deflection of the needle is chiefly due to a difference in temperature of the two parts examined, we must try two new points; and so on, until we find two points of sufficiently equal temperatures.

The second method of testing the equality of the piles by means of the body is as follows :

The piles are tied together, side by side, so as to bring them into close proximity with each other, taking care to have their faces level with each other. They are then enveloped in layers of cotton wool and flannel, leaving only their faces exposed, and the bundle thus constituted is pushed into a wooden shell, open at both ends, until the faces of the piles are 15 mm. (0·59 inch) distant from the end of the shell opposite to that at which they were introduced. This shell has a length of 70 mm. (2·75 inches), and its transverse section measures 50 mm. (1·96 inch) by 35 mm. (1·37 inch). The shell thus enclosing the piles is next wrapped around with thick layers of flannel, and the end nearest to the faces of the piles is applied to the front of the thigh, which is then carefully covered. The faces of the piles are thus enclosed in a small chamber, the floor of which is formed by the surface of the thigh, which the piles do not, however, touch. The temperature of such a chamber may rise to $35°$ C. ($95°$ F.), at the end of twenty minutes, when the temperature of the surrounding atmosphere is $16°$ C. ($60·8°$ F.).

This method possesses the advantage over the preceding one, that it is not necessary to search for two points of equal temperature.

In testing piles for equality in the early stage of an observation, before the reverse currents have set in, we simply shorten the time occupied in the method of procedure given above for testing by direct application to the thigh. The second method of testing cannot be employed for the first phase of an observation, as the time required to warm the enclosure of the wooden shell is too long, and allows the development of the secondary currents to to take place.

CHAPTER II.

METHODS OF TESTING THE RELATIVE TEMPERATURES OF TWO PARTS.—DIFFERENT MOVEMENTS OF THE GALVANOMETER NEEDLES, AND THEIR CAUSES.—COMPOSITION AND CONSTRUCTION OF THERMO-PILES EMPLOYED.—RHOESTAT AND KEYS.—GALVANOMETERS.

Our piles having been tested, let us suppose an experiment made to determine the difference of temperature between two points on the surface of the body, employing the first of the two methods of observation mentioned on p. 8, namely, that of taking the deflection of the needle before the reverse currents commence.

Having the proper resistance for this phase of observation introduced into the circuit of the stronger pile, and having seen that the needle is at zero, when the general circuit is closed, we open the circuit, apply the piles, and immediately close the circuit again, and watch the movement of the needle. If the first swing or impulsion of the needle is great, it is better to stop the experiment at this point, and to diminish the delicacy of the galvanometer by reducing the degree of astasy of the needles,* or by shunting a portion of the combined currents,—otherwise, the needle may be so long in coming to rest that the reverse currents will have time to establish themselves. The delicacy of the galvanometer having been reduced, we proceed as before, and note the degree of deflection when the needle first becomes stationary.

Next, suppose that we adopt the second method, namely, that of waiting until the reverse currents have attained their maximum.

Having introduced the proper resistance for this stage of obser-

---

* It is here supposed that the galvanometer is provided with a controlling magnet.

vation, and supposing the reverse currents not to bear the same proportions to their respective currents proper, we observe the following phenomena on closing the circuit immediately after applying the piles.

1st. The needle may be deflected, in the first moments of the observation to the side opposite to that on which it finally takes up its permanent position; that is to say, the first deflection may indicate the *cooler* of the two parts under examination as the *warmer*. How this may occur will be understood by supposing a case like the following:—

One pile, "a," gives a current proper equal to 8, and a secondary current equal to 4; while the other pile, "b," gives a current proper equal to 6 and a secondary current equal to 4,— this, when both piles are exposed to the same degree of heat and the permanent thermal condition has been fully established. The effective current of "a" will now be $8-4 = 4$; and that of "b" will be $6-4 = 2$. Increasing the resistance of the circuit of "a" until the latter pile has the same strength as "b," we have $\frac{8}{2} = 4$, as the strength of the current proper of "a," and $\frac{4}{2} = 2$, as the strength of the secondary current of "a;" the effective current of "a" being now, therefore, $4-2 = 2$, the same as the effective current of "b." But in the first stage of the experiment, when the currents proper alone are acting, the effective strengths of current will be, for "a," 4, and for "b," 6; hence, at this time, the current of "b" may predominate although the part to which this pile is applied may be actually cooler than the part in contact with "a,"—the question being simply whether the temperature of the part in contact with "a" be superior to that of the part in contact with "b," in a less proportion than that of 6 to 4, which is the proportion of the current strength of "b" to that of "a." Thus if the temperature of the part to which "a" is applied stand in the proportion to that of the part to which "b" is applied of 5 to 4, the current strength of "a" will, at the start, be, $4 \times 5 = 20$; and the current strength of "b," at the same period will be, $6 \times 4 = 24$, "b" thus predominating; but with the full establishment of the secondary currents, "a" will have a power of $5 \times 2 = 10$; and "b," a power of $4 \times 2 = 8$.

If, under such circumstances as the above, the needle has in the first instance passed over to the side of the scale in favour of the cooler part, it will, sooner or later, commence a retrograde course, attain zero, and pass up on the other side of the scale, where it will finally come to rest. To draw conclusions from its position,—which, unless examined carefully, may often appear to be fixed,—at a period anterior to the establishment of the reverse currents, would be, as we have just seen, productive of great error.

2nd. The needle may be deflected to one side of the scale, in the first instance,—remain stationary for a moment or so, and then continue to advance on this side of the scale, coming to rest finally at a point considerably beyond its first halting place.

This would occur when the proportion between the temperature of the part on which "a" is placed, and that of the part on which "b" is placed, is greater than 6 to 4,—for example, is 7 to 4. In this case the current strength of "a" would be represented by $4 \times 7 = 28$; and the current strength of "b" by $6 \times 4 = 24$,—at the commencement of the experiment,—"a" thus predominating by 4; but at the end of the observation, "a" would be represented by $2 \times 7 = 14$, and "b" by $2 \times 4 = 8$,—"a" now predominating by 6.

3rd. The needle may, in the beginning, stand temporarily at zero, and then move to one side and take up a final position there.

This would occur when the proportion between the temperature of the part to which "a" is applied, and that of the part to which "b" is applied, is 6 to 4. In this case, in the first instance, the powers of "a" and "b" are equal, being, respectively $4 \times 6 = 24$, and $6 \times 4 = 24$; but, with the establishment of the secondary currents, the above values are changed to "a" $= 2 \times 6 = 12$; and "b" $= 2 \times 4 = 8$,—hence a deflection of 4 in favour of "a."

4th. The needle may advance on one side, in the first instance, —come temporarily to rest, and then fall back to zero and remain there permanently.

This would occur when the two parts had the same temperature; the first deflection being on the side of "b" as this pile has the stronger initial current.

If proper precautions be used to guard against external disturbing influences, the above four sets of cases will be found to cover all the variations of the movements of the needle ordinarily met with.

Instead of testing the difference of temperature of two parts by placing a separate pile upon each, we may apply one and the same pile, in turn, to each part, having a second pile exposed to a constant source of heat or cold sufficient to keep the needle within proper bounds on the scale. Various forms of apparatus may be used to supply this counter-current, one of which has been described by the author a number of years since;[*] but, if ice cannot be obtained, the body itself furnishes, upon the whole, the most reliable means of supplying this regulating current. The thigh is the part usually selected,—the pile being secured to the limb with tapes, and the part carefully protected from the air. The individual upon whose thigh the pile is placed should have been perfectly quiet for some time (half an hour) previous to the experiment.

There is still another way of testing the comparative temperature of different parts with a single pile, which may be employed with the Thomson or with the author's galvanometer. If the reflected spot of light of these instruments be brought to one end of the scale, we have a range of 720 degrees (scale degrees) from one end of the scale to the other. If the highest temperature to be examined be 36° C. (96·8° F.), and the temperature of the air be 16° C. (60·8° F.), the 720 degrees of the scale of the galvanometer may be made, by regulating the delicacy of the instrument, to correspond to the 20° C. (36° F.) difference between the temperature of the body and that of the air,—that is to say, 36 degrees of the scale of the galvanometer will correspond to 1° C.,—this, of course, measuring roughly. This single pile is placed in turn on each of the points to be examined.

We will next examine the composition and construction of the particular piles employed in the experiments related in this work.

The piles have, in the greater number of cases, consisted of one of two combinations, namely,—bismuth and certain alloys of antimony; and German-silver and iron. The antimony alloys employed, have been chiefly of three kinds. The following are the compositions in 100 parts by weight of each of these alloys:

	No. 1.	
Antimony		59·798
Cadmium		25·730
Zinc		14·472
	No. 2.	
Antimony		40·464
Cadmium		37·673
Zinc		21·863

---

[*] 'Archives de Physiologie,' July—August, 1868, t. 1, p. 498.

No. 3.
Antimony . . . . 64·429
Zinc . . . . . 35·571

Decided differences of electro-motive force are found in different specimens of these alloys, although prepared in apparently the same way. This appears to be especially the case with the alloys containing cadmium. Schiff, noticing this difficulty in certain alloys of antimony employed by him, has attributed it to either the degree of heat employed in the fusion of the metals, or to the rapidity of the subsequent cooling.*

Whatever the cause may be, it does not appear to be readily avoidable, as it is no uncommon thing to find bars of the above alloys, which were cast at one and the same time, of very different electro-motive forces. When possessed of their full powers, No. 1 of the above alloys has an electro-motive force of about 6 times, and No. 3 of about 3·4 times, that of pure antimony. No. 2 possesses an electro-motive force intermediate in degree between the two preceding values.†

The alloys and bismuth are cast in bars 19 mm. (0·74 inch) in length. The exposed junction of each pair is 2 mm. (0·078 inch) square. The piles are constructed of from one to eight pairs. The construction of those of a single pair is as follows :—

The pair is surrounded, in its entire length, with a coating of paraffin 2 mm. (0·078 inch) deep, and the farther ends of the bars, to which are joined the conducting wires, are covered with a layer of the above substance 4 mm. (0·157 inch) thick. The pair thus protected is enclosed in an ebonite cylinder, the sides of which are 1 mm. in thickness, closed at one end, where it is pierced with two holes for the passage of two straight copper wires 40 mm. (1·57 inch) in length, and of a diameter of 0·761 mm. (0·03 inch), forming the commencements of the conducting wires. The open end of the cylinder leaves the junction of the pair exposed ; and the entire surface of this end,—walls of cylinder, paraffin, and ends of bars,—is made flush, so that, when applied to the surface of the body, the whole pile presses, at every point, evenly and equally upon the part examined. The exposed junction is then varnished.

* 'Archives de Physiologie,' t. 11, p. 164, 1869.
† The author is indebted for a knowledge of the thermo-electric powers of these alloys, which he has employed since 1866, to Prof. Moses G. Farmer, of the United States Navy Torpedo Station, Newport, Rhode Island.

To the two straight copper wires connected with the upper ends of the bars, are joined two conductors consisting of silk-covered strands of copper wire, each conductor being 1000 mm. in length; and to the distant end of each conductor is attached a copper wire similar to that forming the central termination of the conductor. The straight copper terminals of the pile are each thickly wrapped in cotton wool, soaked afterwards in paraffin, commencing at the points where they emerge from the top of the ebonite casing, and extending to a distance of 30 mm. (1·18 inch) beyond their attachments to the strands. The two wires thus protected are tied together and further covered with flannel, which extends down for a couple of millimetres over the upper end of the ebonite casing. It will be seen that all parts of the pile except the face, are well protected from external variations of temperature, as are also the heterogeneous points where the strands join the straight wires. A little beyond the coverings, the conducting strands are bent over in a loop and made fast to the bundle formed of the straight copper terminals. The object of this is to prevent vibrations of the strands from propagating themselves onward to heterogenous points, thus giving rise to thermo-electric currents. Piles of two or more pairs are constructed on the same plan.

German-silver and iron elements have been employed only in cases where it was necessary to give to the bars a very small sectional area, and also to have the pair protrude for a little distance,—5 mm. (0·196 inch),—beyond the ebonite casing, in order to examine the skin beneath thick hair. The portion of the bars protruding from the ebonite is covered with sealing wax so as to give to this end of the pile a pencil shape. In other respects, the pile is constructed similarly to that already described.

The piles are held in brass clamps attached to the ends of rods of the same metal, which slide in and out of wooden handles. In applying the piles to the skin, great care must be used to make the pressure equal in each point of the part tested. One should attend carefully that the outer edge of the ebonite casing, or of the envelope of sealing wax, be everywhere in close contact with the surface. In careless or inexperienced hands, it is common to find only a part of the face of the pile in close contact with the skin, the outer part being tilted up from the surface by the unevenness of the pressure applied.

With piles constructed like those first described, the superficial vessels can be emptied by firm pressure, leaving the surface bloodless on removing the piles.* With such piles we may test the temperature of parts lying directly beneath the temporal artery, without the blood of the latter interfering; the flat end of the pile emptying the vessel, and the paraffin packing, lying between the ends of the bars and the edge of the ebonite casing, preventing the temperature of the blood outside the compressed area from affecting the junction of the pile by conduction.

The conducting wires of the pile pass to an instrument which unites in one, keys for four currents, a rheostat, and a commutator, —all acting through the medium of mercury cups. This instrument in its earlier form has already been described some years since;† in its present shape, however, it has undergone modifications which render a new description of it necessary. Fig. 1 gives the general plan of the instrument.

E, E, E, E, is a slab of ebonite, 200 mm. (7·87 inches) in length, by 167 mm. (6·57 inches) in width, and 11 mm. (4·43 inches) in thickness; this slab forming the top of a wooden box, 40 mm. (1·57 inch) in depth. B^1, B^2, B^3, B^4, are four sets of binding-screws, all the left hand ones of which, l^1, l^2, l^3, l^4, are connected together and with the mercury cup m^{13}, by the copper bands w, w, &c. The right hand binding screws, r^1, r^2, r^3, r^4, are connected severally by the bands, x, x, x, x, with the mercury cups, a^1, a^2, a^3, a^4. In front of each of these latter are situated two cups, $b^1 c^1, b^2 c^2, b^3 c^3, b^4 c^4$. The cups, b^1, b^2, b^3, b^4, are all connected together by the bands y, y, &c., and are also connected with mercury cups, m^1 and m^{11}. The cups, c^1, c^2, c^3, c^4, are likewise in communication with each other and with mercury cup m^{10} through the copper bands z, z, z, &c. It will be observed that the connections, z, z, z, &c., pass under, without communicating with, the connections y, y, y, &c. The same is the case with w, w, w, &c., with reference to x, x, x.

The mercury cups m^1 to m^{10}, both included, form the connections of the rheostat portion of the instrument. Nine wires of different resistance, wound upon as many spools placed in the box beneath the ebonite slab, communicate each with a pair of mercury cups. The wire on the first spool has one end in cup m^1

---

* This emptying of the superficial vessels is the second of the two causes of the retrograde movement of the needle, mentioned on page 6.

† 'Archives de Physiologie,' t. 1, p. 498, July and August, 1868.

Fig 1.

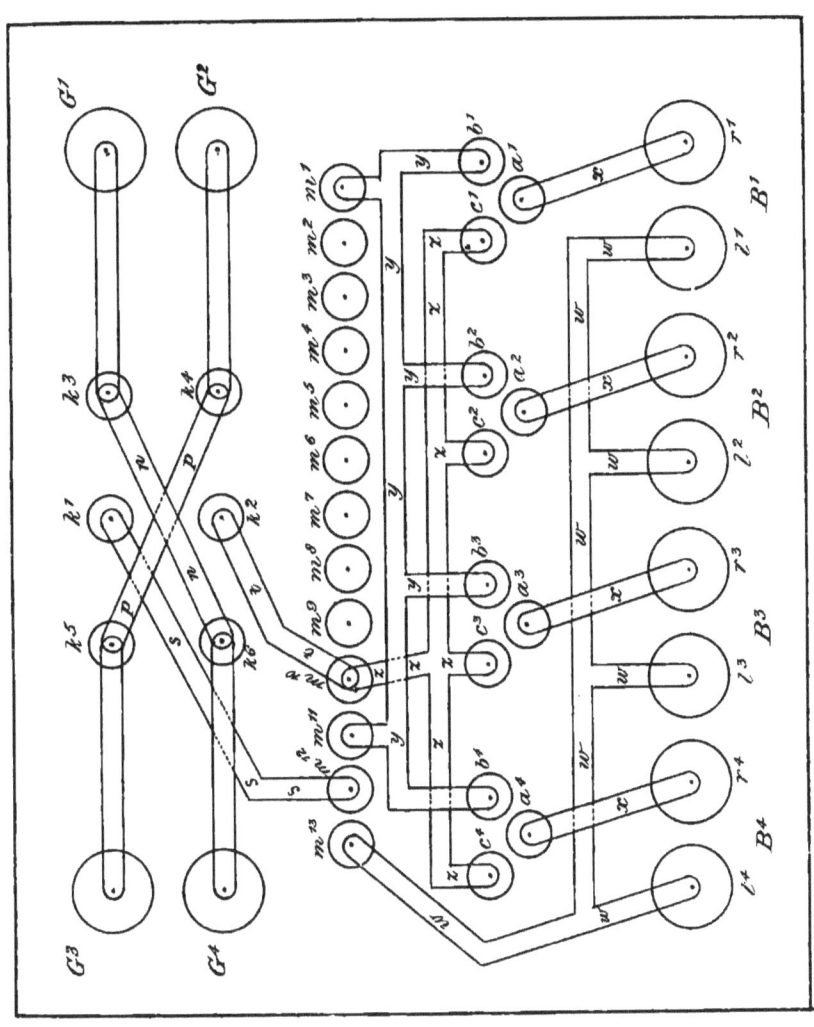

and the other end in cup m^2; the second resistance coil has one end in m^2 and the other in m^3, and so on through the row of cups to m^{10}. The cups are covered over with a strip of ebonite hollowed out beneath so as to form a sort of arch above them. Nine copper forks, the prongs of which dip into the mercury cups, connect the individual cups with each other. The handles of these forks pass through holes in the ebonite roof of the cups, working up and down with friction. These handles are provided at the top with ebonite knobs. If now the fork connecting a pair of cups be raised, the direct connection between the two cups being thus interrupted, a current arriving at one cup will be obliged to pass through the resistance coil, the ends of which are in communication with the pair of cups. We can, thus, by raising the forks, cause the current to pass through as many of the resistance coils as we may see fit.

The cups m^{10} and m^{11} are also provided with a connecting fork as are likewise cups m^{12} and m^{13}:—the use of these last two sets of cups will be presently shown.

The four sets of cups $a^1\ b^1\ c^1$, $a^2\ b^2\ c^2$, $a^3\ b^3\ c^3$, $a^4\ b^4\ c^4$, are covered in like the preceding row of cups, and each set is provided with a fork. The fork of each set of cups admits of two separate connections,—it may unite " a " to either " b " or " c," according as one prong is dipped into " b " or " c " respectively, the other prong remaining, in both instances, in " a." Moreover, when the fork is raised above the level of the cups it acts as a key to the particular set of binding screws the right hand one of which it is in communication with.

The mercury cups k^1, k^2, k^3, k^4, k^5, k^6, with the permanent cross piece " p," and the movable cross piece " n " form the base of the commutator, which is the well-known wippe, or rheotrope of Pohl. The binding screws $G^1\ G^2$ and $G^3\ G^4$, are connected with galvanometers or other apparatus. It will be observed that k^1 communicates through the copper band " s " with cup m^{12}, which when connected with m^{13} by the fork belonging to the pair, communicates, in turn, with all the left hand binding screws of the sets B^1, &c., k^1 is, therefore, one terminating point of all currents entering or leaving the sets of binding screws B^1, &c., and the fork between m^{12} and m^{13} forms a general key, cutting off all currents, at will. In like manner, k^2 forms the other terminating point of all currents entering or leaving the sets of binding screws B^1, &c. k^2 is united with m^{10} by the band "t," and m^{10} is in communication through three different paths with all the right hand binding screws.

Let us now suppose a thermo-pile connected with the two binding screws of the set B^1, the current entering at "r." We will suppose a^1 and b^1 to be connected; m^{12} and m^{13} to be likewise connected; m^{10} and m^{11} to be disconnected; the rocker of the rheotrope to be thrown over to the side of k^3 k^4, and G^1 G^2 to be in communication with a galvanometer.

The current entering at "r" passes over the fork connecting a^1 and b^1 and arrives at the cup m^1. It is now in our power to impose upon the current what resistance we may choose in its passage to m^{10}.* Arrived at this last point it passes on to k^2, thence through the rocker of the rheotrope to k^4 and G^2, and through the galvanometer,—returning through G^1, k^3, k^1, "s," m^{12} m^{13}, and thence through "w" to l^1. We have supposed m^{10} and m^{11} to be disconnected:—if now after having raised a number of the forks of the rheostat, we wish at once to throw off all resistance, instead of severally lowing each fork in turn, we have only to connect m^{10} and m^{11}, and a direct path outside of the rheostat is secured to the current. If "a" and "c" be in communication, it will be seen that the current reaches m^{10} without passing through the rheostat. These two separate paths from "a" to m^{10} the one through, and the other exterior to, the rheostat permit us, when employing two piles, to weaken the current of one without affecting that of the other, as is necessary in the case where one pile has a greater electro-motive force than the other. Let us suppose two piles acting against each other, of which one, F, is stronger than the other, H. F has its conducting wires attached to the set B^1 of binding screws, its positive pole being at r^1, while H communicates with the set B^2, its positive pole being at l^2. In the circuit of F, a^1 is connected with b^1,—m^{10} and m^{11} being disconnected,—so that the current is obliged to pass through the rheostat, where it is weakened to the point of equality with the current of H. This latter instead of passing by the route a^2 b^2, y, takes that of a^2 c^2, z, on its return from the galvanometer,

* The resistances usually employed have been as follows:—0·00625 ohm, 0·0125 ohm, 0·025 ohm, 0·05 ohm, 0·1 ohm, 0·2 ohm, 0·4 ohm, 0·5 ohm, 0·8 ohm. Thus admitting of an increase of 0·00625 ohm from that sum up to a total of 2·09375 ohm.

thus avoiding the rheostat; both currents can, of course, be cut off at m^{13} m^{12}.

All connections between binding screws and conducting wires should be thoroughly enveloped in cotton wool, to guard against changes of temperature from without affecting these points. The rheotrope must also for the same reason be well protected. Moreover, the wires from the pile must be made fast to the table on which the above instrument is placed, at a point anterior to the junction of the strands with the copper terminals, so that vibrations of the conductors of the pile will be checked before reaching the junction in question, or, further still, the points of contact of the copper terminals with the binding screws.

The galvanometer employed next claims our attention. Both Sir W. Thomson's galvanometer, and an instrument devised by the author have been used. The latter is an improved form of an instrument described in 1875.* It admits of both direct reading of the deflection of the needles, and of indirect reading by a reflected ray of light from a mirror, as in the Thomson galvanometer. The direct reading is made by means of an aluminium index 60 mm. (2·36 inches) long, placed at right angles to the upper of the two needles of the astatic system, the index ending in a fine point which moves over a graduated arc. The reflecting mirror is only 5 mm. (0·196 inch) in diameter. The reflection may be employed in one of two ways, namely:

1st. The light is concentrated upon the mirror by means of a lens, and a circular spot of light is thrown upon the scale placed at about 1000 mm. (39·37 inches) distance, this spot being cut by a fine blackened wire, serving as an index. The scale in this case is similar to that of Sir W. Thomson.

2nd. A scale, one division of which corresponds to three divisions of the former scale, is placed at a distance of 3000 mm. (118·11 inches) from the mirror. One degree of the latter scale is therefore equivalent to one degree of the Thomson scale at 1000 mm. distance. The larger scale which is one metre long by 130 mm. (5·11 inches) wide is fastened on the inside of the bottom of a large deal box placed upon its side. The reflection shows itself by a circular spot of light 130 mm. in diameter, cut, as before, by the defining wire. Using this scale, one may easily read deflections at a distance of four metres (157·48 inches).

* 'British Medical Journal,' January 23rd, 1875.

Temperature of the Head.

The galvanometer is provided with two adjusting magnets. It has also two shunts and a key, all acting through mercury cups. Fig. 2 represents the plan of the key and shunts, full size. B and C are two binding screws connected with the poles of the

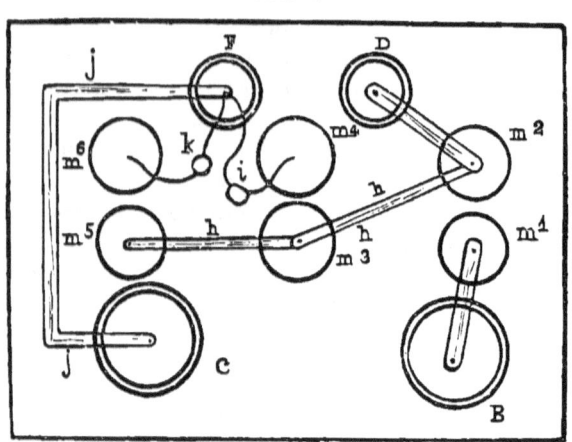

FIG. 2.

pile; D and F are the binding screws connected with the galvanometer.* A current entering at B arrives at the mercury cup m^1 which with m^2 forms the key:—the connecting fork between the two cups being depressed the current passes on to D, traverses the coil of the galvanometer and returns through F, j, to C. m^2 is connected by h, h, with mercury cups m^3 and m^5; and mercury cups m^4 and m^6 are connected, respectively, by coils of wire i and k with F. Suppose now m^3 and m^4 to be connected, a portion of the current arriving in m^2 will now be diverted through h, m^3, m^4, i, F, and so back through j to C. m^5 and m^6 act in a similar manner. The shunts usually halve and quarter, respectively, the deflection of the galvanometer.

* When the galvanometer is used for medical purposes (in which case the mirror is unnecessary), with a given pair of piles,—instead of having two shunts, only one is employed, and two resistances are added for the stronger pile, one resistance being for the first and the other for the second phase of observation.

All the instruments described in this work are made by Mr. Adam Hilger, 192, Tottenham Court Road, London, to whose skill and intelligence the author is much indebted.

PART II.

THE RELATIVE TEMPERATURES OF DIFFERENT PARTS OF THE SURFACE OF THE HEAD IN THE QUIESCENT MENTAL STATE.

CHAPTER I.

GENERAL REMARKS—DIVISIONS AND MEASUREMENTS OF THE HEAD.—SUBDIVISIONS AND MEASUREMENTS OF THE ANTERIOR REGION.—EXAMINATION OF THE ANTERIOR REGION IN SYMMETRICALLY SITUATED SPACES OF THE TWO SIDES.

THE first step to be taken having been decided upon, namely,—a thorough examination of the relative temperatures of small subdivisions of the surface of the head, the next point demanding attention was the particular manner in which the examination could be best carried out. Preliminary observations had satisfied the author that to obtain accurate results, the method to be adopted was not to examine a great number of heads taken at random, but to limit the observations to a thorough and minute, examination of a small number of heads, which could be accurately measured, and compared one with another, and in which the normal variations of temperature, and the effect of different mental and physical conditions, could be carefully studied.

In the first place, with reference to the measurements of the heads, it is indispensably necessary, that the surface be mapped out in determinate districts, measured,—and *marked*, where it is possible,—with care, and that these districts occupy the same anatomical positions in the different heads. If this be not accurately attended to, serious errors are certain to occur. The comparative superiority of temperature in the head often shifts its position from one side to the other, and from one point to another, on one and the same side, within such narrow limits, that the moving of a thermo-pile 5 millimetres (0·196 inch) may entirely reverse the position of higher temperature previously found. All such broad distinctions as temporal, superciliary, &c., and rough measurements made with the unaided eye, are almost sure to lead to confusing and contradictory results.

Moreover, it is necessary to know as much as possible concerning the various circumstances both internal and external affecting the subjects of examination; the general temperament, mental condition, occupation,—more especially what it has been just before the examination; the state of the circulation; the temperature of the atmosphere to which the individuals have been exposed before the experiment; whether they have been near a fire; exposed to the rays of the sun, or to drafts of air, &c.; and if exposed to any of the above influences, whether one part of the head was acted upon more than another; also the previous position of the head and body should be known,—whether one side of the head has been resting on a pillow, back of a chair, &c. All these points must be attended to in order to ensure accuracy. In another place the variations from the normal temperature brought about by the conditions and influences enumerated above will be taken into consideration; they are alluded to here merely to indicate the importance of a thorough acquaintance with every possible detail of both the internal and external circumstances in which the individuals experimented on are placed.

The number of heads strictly comparable which have been examined has been limited to six,—three males and three females. As, however, one of the individuals was examined only a few times, and in a limited area, the main part of the results were obtained on five heads,—three females and two males.

The first subject which claims attention is the division of the surface of the head into determinate districts.

Divisions of the surface of the head.

The surface of the head is divided into three regions, designated, respectively, *anterior*, *middle*, and *posterior*.

Anterior region.

The anterior region is bounded laterally by a line drawn upward, on each side of the head, from the angle formed by the

Divisions of Surface of the Head.

frontal and zygomatic processes of the malar bone, in a direction parallel to the plane of the forehead taken over the frontal eminences and the superciliary ridges.

The superior boundary is formed by the continuation, in the same plane, and junction on the top of the head, of the lateral boundaries.

The inferior boundary is formed by a line passing horizontally across the front of the head on a level with the summits of the supra-orbital arches, between the external angular processes of the two sides, and thence continued by the outer borders of the malar bones to the points of origin of the lateral boundaries.

The part of the head embraced by the anterior region is, therefore, that which would be cut off anteriorly by a transverse and vertical section, made parallel to the plane specified, between the angles formed by the frontal and zygomatic processes of the malar bones of the two sides.

Middle region.

The middle region is bounded anteriorly by the lateral and superior boundaries of the anterior region.

The posterior boundary is formed by a line passing over the top of the head, parallel to the anterior boundary, uniting the extremities of the mastoid processes of the two sides.

The inferior boundaries, commencing at the terminations of the inferior boundary of the anterior region follow the upper borders of the zygomatic processes of the temporal bones, pass behind the ears, and follow the anterior borders of the mastoid processes to their extremities.

Posterior region.

The posterior region has for its superior and lateral boundaries the posterior boundary of the middle region.

The inferior boundary is formed by the posterior borders of the mastoid processes and by the superior curved line of the occipital bone.

The longitudinal medium line of the head divides each of the three regions into right and left symmetrical halves.

It will be seen that all the boundaries of all three regions, except the inferior boundaries, depend for their line of direction upon the direction of the plane of the forehead. It is obvious that this mode of measurement would be likely in different heads to enclose tracts not strictly comparable. It was chosen in the present investigations simply for the reason that it appeared to be better suited to the particular heads examined than any other mode of measurement that suggested itself to the author. Every head was referred to one head as a standard; and the measurements of the different regions and subdivisions of the standard head, with certain landmarks laid down in each region, render it not difficult to apply the results given in this work to other heads.

We will proceed first to examine the anterior region in detail.

The anterior region is thus subdivided:—Commencing at the inferior boundary each lateral half is divided into six parts by five equidistant lines drawn horizontally from the median line to the lateral limit. The tracts thus marked off are designated *tiers*, and are numbered from 1 to 6, commencing at the inferior boundary. Further in each lateral half four equidistant vertical lines are drawn upward, parallel to the median line, from the inferior boundary to the superior boundary, thus dividing each tier into smaller spaces. In the first four tiers these spaces are five in number; but in the fifth and sixth tiers the convergence, over the top of the head, of the lateral limits, diminishes the length of the tier, and reduces the number of spaces to four in the fifth tier and to three in the sixth tier. These vertical subdivisions are called *districts*, and are numbered from 1 to 5 from the median line outward.

The tiers and districts are marked off by means of a string coated with coloured chalk. All the spaces are thus sharply and evenly defined.

The following are the measurements of the anterior region of the standard head.

Distance between inferior and superior boundaries, measured on the median line, 125 mm. (4·92 inches). Total breadth, from side to side, measured on the horizontal portion of the inferior boundary between two points on the same vertical line with the terminations of the inferior boundary, 186 mm. (7·?2 inches). Total breadth, measured between lateral boundaries over middle of frontal eminences, 186 mm. (7·32 inches). Total breadth, measured between lateral boundaries just above frontal eminences, 176 mm. (6·928 inches). Total breadth, measured across the head at a distance, on the median line, of 20·83 mm. (0·82° inch) from

Subdivisions of Anterior Region. 31

superior boundary, 120 mm. (4·72 inches). The distance over the top of the head, following the line of the lateral and superior boundaries, between the terminations of the inferior boundary, is 319 mm. (12·559 inches).

The height of the anterior region on the median line being 125 mm. (4·92 inches), each of the six tiers will measure vertically, on the median line, 20·83 mm. (0·82 inch). The upper boundary of the 1st tier touches the summits of the superciliary ridges. The upper boundary of the 2nd tier passes through the centre of the frontal eminences (line of second measurement of breadth). The upper boundary of the 3rd tier touches the upper border of the frontal eminences (line of third measurement of breadth). The upper boundaries of the 4th and 5th tiers have no anatomical landmarks, and their positions can only be designated by their respective distances from the upper boundary of the 3rd tier, or from the superior boundary of the region, which latter limit touches the coronal suture on the median line. The distance of the upper boundary of the 4th tier from the superior limit of the region is 41·66 mm. (1·64 inch); and the distance of the upper boundary of the 5th tier from the superior boundary of the region is, of course, one half the last amount, namely, 20·83 mm. (0·82 inch).

The breadth of each lateral half, measured on the inferior boundary line, being 93 mm. (3·66 inches), each district will measure horizontally on this line 18·6 mm. (0·732 inch). The outer boundary of the 1st district is on a line with the inner border of the supra-orbital notch. The outer third of the 4th district is in a line with the external angular process.

Bearing in mind the size of the faces of the pairs forming the thermo-piles, it will be evident that the size of the subdivisions of the surface of the anterior region permits of each space being tested, with the smaller piles, in several distinct parts. When spaces are thus examined, the average of the different examinations is taken to represent the whole space. It will be seen further on, that the subdivisions of the middle and posterior regions are also of a size to permit of a similarly minute examination.

We have 27 spaces on each side, to examine in the anterior region.

Temperature of the Head.

Comparison of symmetrically situated spaces of the two sides of the anterior region.

TABLE I.—Results of 100 observations on the comparative temperature of each pair of symmetrically situated spaces of the two sides of the anterior region. "R" signifies right side; "L," left side; and "N," equality of the two sides. The figures prefixed to "L," "R," and "N," denote, respectively, the number of times in a total of 100 in which the temperature was higher on the right side, the left side, or equal on the two sides.

	1st district.	2nd district.	3rd district.	4th district.	5th district.
6th Tier.	32 L. 68 R.	29 L. 71 R.	38 L. 62 R.		
5th Tier.	30 L. 65 R. 5 N.	26 L. 70 R. 4 N.	30 L. 65 R. 5 N.	28 L. 69 R. 3 N.	
4th Tier.	32 L. 55 R. 13 N.	32 L. 44 R. 24 N.	30 L. 40 R. 30 N.	23 L. 65 R. 12 N.	32 L. 68 R.
3rd Tier.	60 L. 18 R. 22 N.	64 L. 12 R. 24 N.	75 L. 15 R. 10 N.	28 L. 60 R. 12 N.	30 L. 70 R.
2nd Tier.	74 L. 26 R.	76 L. 24 R.	58 L. 30 R. 12 N.	18 L. 60 R. 22 N.	24 L. 76 R.
1st Tier.	74 L. 20 R. 6 N.	77 L. 23 R.	50 L. 34 R. 16 N.	35 L. 65 R.	32 L. 68 R.

The first thing to which attention is called in Table I is, *that in no one of the subdivisions of the anterior region, is the temperature uniformly higher on one side than on the other; on the contrary in every space it may be higher on the right side or on the left side, in turn.*

We have, therefore, in each pair of symmetrically situated spaces, to consider merely on which side of the head, *in the majority of cases*, the higher temperature is found. We leave out, for the moment, the cases of equality of temperature.

Commencing with the 1st tier, we have for each district of each tier the following proportions in numbers of times of superiority of temperature on the right and left sides, respectively:—

1st Tier.

1st District—3·7 times to 1 in favour of left side
2nd   „   3·34   „   „
3rd   „   1·47   „   „
4th   „   1·85   „   „   right side.
5th   „   2·12   „   „

2nd Tier.

1st District—2·84 times to 1 in favour of left side.
2nd   „   3·16   „   „
3rd   „   1·93   „   „
4th   „   3·33   „   „   right side.
5th   „   3·12   „   „

3rd Tier.

1st District—3·33 times to 1 in favour of left side.
2nd   „   5·33   „   „
3rd   „   5·00   „   „
4th   „   2·14   „   „   right side.
5th   „   2·33   „   „

4th Tier.

1st District—1·72 times to 1 in favour of right side.
2nd   „   1·37   „   „
3rd   „   1·33   „   „
4th   „   2·82   „   „
5th   „   2·12   „   „

5th Tier.

1st District—2·17 times to 1 in favour of right side.
2nd   „   2·69   „   „
3rd   „   2·17   „   „
4th   „   2·46   „   „

6th Tier.

1st District—2·12 times to 1 in favour of right side.
2nd   ,,   2·45   ,,   ,,
3rd   ,,   1·63   ,,   ,,

Thus in the twenty-seven spaces compared, the average relative temperature is found to be higher on the right side than on the left side in eighteen spaces, or two thirds of the whole number.

Diagram 1 gives a general idea of the distribution of temperature in the anterior region. The 1st district, 1st tier, is of doubtful value, of course, with regard to the temperature of the brain, on account of the frontal sinuses. Only the upper parts of the 2nd and 3rd districts of the same tier are likewise available for examination, on account of the orbit.

If we take the total number of observations in which either the right or the left side was the warmer, we find that out of this total, namely, 2480, the right side was superior in temperature in 1343 instances, and the left side in 1137 instances. The percentages of superiority of temperature for the right and left sides, respectively, are, therefore, 54·153, and 45·847.

But if we take the averages of all the proportionate numbers of times in which each side was superior in temperature to the other, we find that the left side has the greater average; the mean for the nine spaces in which this latter side has a majority, being 75·069 per cent., while the mean for the eighteen spaces in which the right side has a majority is 68·117 per cent.;—that is to say, —in those spaces in which the left side has a majority of cases of higher temperature, this majority is, on an average, greater than the average majority found in the spaces in which the right side has the larger number of cases of superior temperature.

We pass next to those cases in which the temperature is equal on the two sides. Referring to table 1, the following is found to be the distribution of the neutral points, and the proportionate numbers of times of their occurrence.

1st Tier.

1st District—1 in 16·666 times
3rd   ,,   1 in 6·25   ,,

DIAGRAM 1.

Distribution of comparative superiority of temperature on the two sides of the anterior region.

The shaded spaces are those which, of symmetrically situated spaces on the two sides, have the higher temperature. "Right" and "Left" signify respectively right and left sides of the head.

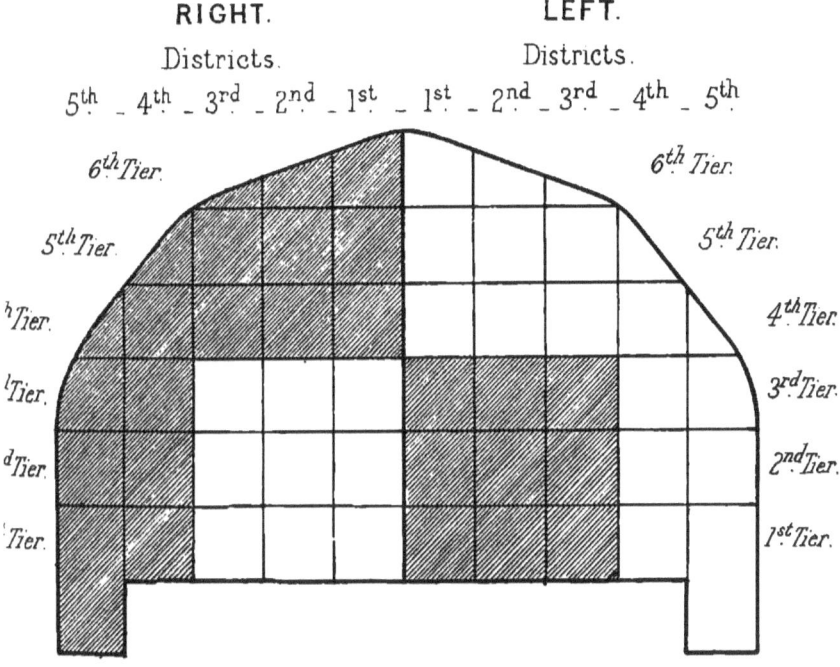

Examination of Anterior Region.

2nd Tier.
3rd District—1 in 8·333 times.
4th   ,,   1 ,, 4·555   ,,

3rd Tier.
1st District—1 in 4·545 times.
2nd   ,,   1 ,, 4·166   ,,
3rd   ,,   1 ,, 10·000   ,,
4th   ,,   1 ,, 8·333   ,,

4th Tier.
1st District—1 in 7·692 times.
2nd   ,,   1 ,, 4·166   ,,
3rd   ,,   1 ,, 3·333   ,,
4th   ,,   1 ,, 8·333   ,,

5th Tier.
1st District—1 in 20·000 times.
2nd   ,,   1 ,, 25·000   ,,
3rd   ,,   1 ,, 20·000   ,,
4th   ,,   1 ,, 33·333   ,,

Thus equality of temperature of the two sides is found in sixteen spaces, or in 59·259 per cent. of the whole number of spaces. Of the total number of observations contained in table 1, namely, 2700,—220, or 8·148 per cent., show equality of temperature. The mean percentage of cases of equality of temperature in the total number of observations in the sixteen spaces in which equality is found, is, 13·75.

In the next place taking the total number of observations,—2700,—we have the following percentages of times of occurrence of superiority of temperature on the right and on the left sides, and of equality of temperature on the two sides:—

Right side—50·112 per cent.
Left   ,,   41·740   ,,
Neutral   8·148   ,,

But in eleven of the twenty seven spaces equality of temperature is not found. If, therefore, we take only the sixteen spaces in which all three conditions exist we have another set of percentages :

36   *Temperature of the Head.*

>   Right side—45·125 per cent.
>   Left   „   41·125   „
>   Neutral   13·750   „

Here the percentage for the left side is very slightly diminished; that for the right side more considerably diminished; while that for equality is decidedly increased.

We will next take the percentages of right and left superiority of temperature, and of equality of temperature for each entire tier, and for each entire district.

Percentages for Tiers.

	Right side.	Left side.	Neutral.
1st Tier	44·00	53·60	2·40
2nd „	43·20	50·00	6·80
3rd „	35·00	51·40	13·60
4th	54·40	29·80	15·80
5th „	67·25	28·50	4·25
6th „	67·00	33·00	0·00

The 5th tier shows the highest percentage for the right side; the 1st tier the highest percentage for the left side; and the 4th tier the highest percentage for neutrality.

Percentages for Districts.

	Right side.	Left side.	Neutral.
1st District	42·000	50·333	7·666
2nd „	40·666	50·666	8·666
3rd „	41·000	46·833	12·167
4th „	63·800	26·400	9·800
5th „	70·500	29·500	0·000

The 5th district shows the highest percentage for the right side; the 2nd district the highest for the left side (the 1st district has very nearly as high a percentage); and the 3rd district the highest for neutrality.

With regard to individual spaces, the following is the distribution and rate of the highest percentages in favour of the right side, left side, and equality, respectively.

In favour of Right Side.

5th District of 2nd Tier—76 per cent.
5th　　,,　　3rd　,,　　70　　,,
2nd　　,,　　5th　,,　　70　　,,
4th　　,,　　5th　,,　　69　　,,
2nd　　,,　　6th　,,　　71　　,,

In favour of Left Side.

1st District of 1st Tier—74 per cent.
2nd　　,,　　1st　,,　　77　　,,
1st　　,,　　2nd　,,　　74　　,,
2nd　　,,　　2nd　,,　　76　　,,
3rd　　,,　　3rd　,,　　75　　,,

In favour of Equality.

4th District of 2nd Tier—22 per cent.
1st　　,,　　3rd　,,　　22　　,,
2nd　　,,　　3rd　,,　　24　　,,
2nd　　,,　　4th　,,　　24　　,,
3rd　　,,　　4th　,,　　30　　,,

To sum up the principal points set forth in table 1,—we see that, starting with the 1st district, 1st tier, the average temperature is higher on the left side than on the right side in the 1st, 2nd, and 3rd districts of the first three tiers; that in the 4th tier, passing upward, and in the 4th district passing outward, the balance of superiority of temperature shifts over to the right side, —so that, while the inner and lower portion of the anterior region is of higher temperature on the left side, the outer and upper portion is of higher temperature on the right side; that neutral points occur most frequently in those tiers and districts in or near which the balance of higher temperature passes over from one side to the other.

It has been stated before (p. 31) that the size of the faces of some of the piles permits of the examination of several points in one and the same space, and that the figures set down in the tables for a given space are deduced from the total of the different observations made in that space. Now in one and the same space different points may furnish different results; thus, the upper part of a space may show a greater majority of cases in favour of a given side than the lower part of the same space. For instance, in a certain space we may obtain, in twenty-five examinations of its upper part, twenty results in favour of the right side and five results in favour of the left side; while, in an equal number of examinations of the lower part of the same space we may obtain fifteen results in favour of the right side and ten results in favour of the left side. These results would be estimated in the table as thirty-five right and fifteen left, for the whole space. Or we might find in the upper part of the space, twenty results in favour of the right side, and five results in favour of the left side; while in the lower part we might get ten results in favour of the right side and fifteen results in favour of the left side; which results would be tabulated as thirty right and twenty left. Also with regard to equality of temperature, it may occur oftener in one part of a space than in another, or it may be confined to one part. These unequal distributions of temperature in one and the same space, are most frequent near the lines where the balance of higher temperature usually shifts from one side to the other; that is, in the 3rd and 4th districts of the 1st and 2nd tiers, and in the first four districts of the 3rd and 4th tiers. If we examine these border spaces in subdivisions, we usually obtain results, as regards right and left superiority of temperature, such as are mapped out in diagram 2. It will be observed that the distribution of temperature in this last diagram differs somewhat from that set down in diagram 1, in that, in diagram 2, the 1st, 2nd, 3rd, and 4th districts of the 4th tier, and the 4th district of the 2nd and 3rd tiers are not put down as higher in temperature on the right side in their *whole extent*, but only in *two thirds of their extent* in five of the spaces specified, and in *somewhat more than two thirds* (seven ninths) in the remaining space,—4th district, 4th tier.

But although the results which have been given cover the majority of cases, yet anomalies are frequent, even when the utmost care is used to avoid all disturbing influences. Two of the most common of these anomalies are show in diagrams 3 and 4. In diagram 3 we see an extension of the region of higher temperature on the right side; and in diagram 4 we see a similar extension on the left side. Moreover, these extensions of the tracts of higher temperature may proceed so far as to cover almost

DIAGRAM 2.

Distribution of comparative superiority of temperature on the two sides of the anterior region. Mean results.

The shaded spaces are those which, of symmetrically situated spaces on the two sides, have the higher temperature. " Right " and " Left " signify respectively right and left sides of the head. Thus, the 5th district, 1st tier, being shaded on the right side and unshaded on the left side, denotes that this space is usually warmer on the right side than on the left.

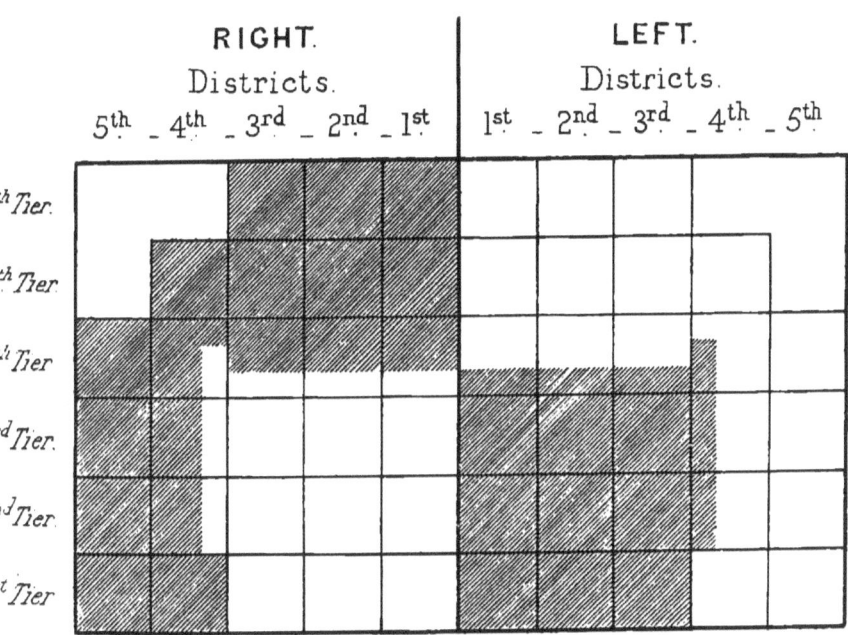

DIAGRAM 3.

Increase in the extent of the tract of higher temperature on the right side of the anterior region.

The shaded spaces are those which, of symmetrically situated spaces of the two sides, have the higher temperature. "Right" and "Left" signify respectively right and left sides of the head.

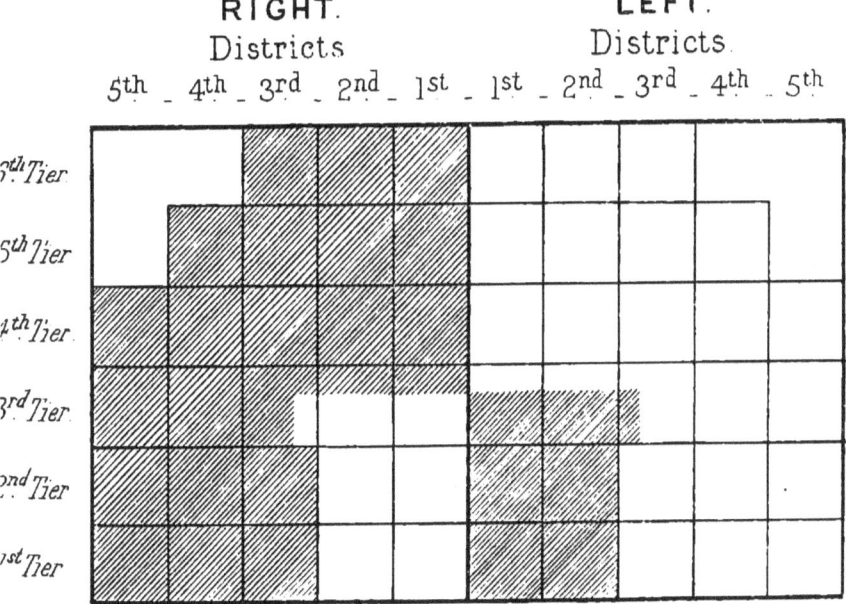

DIAGRAM 4.

Increase in the extent of the tract of higher temperature on the left side of the anterior region.

The shaded spaces are those which, of symmetrically situated spaces of the two sides, have the higher temperature. "Right" and "Left" signify respectively right and left sides of the head.

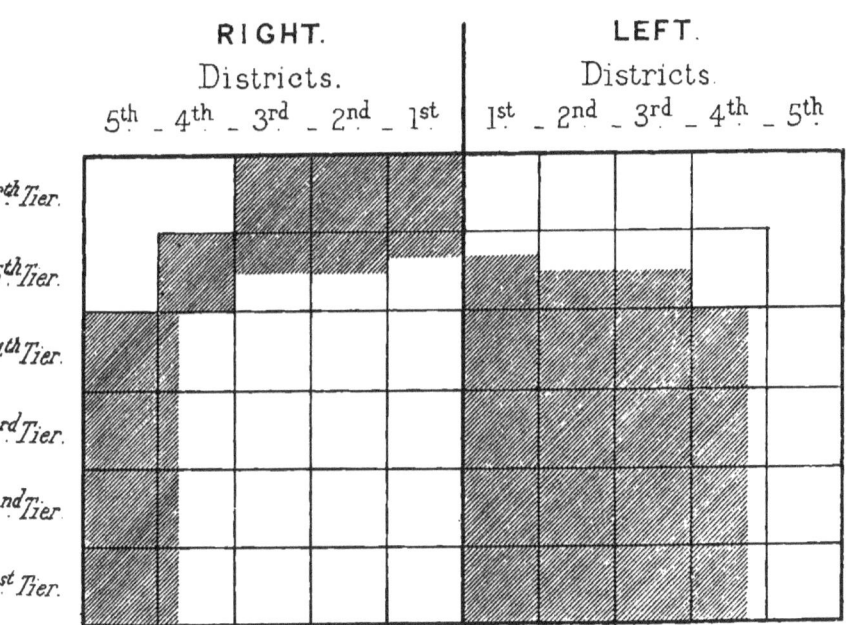

Examination of Anterior Region. 39

the whole of one side of the anterior region, thus making nearly every space warmer on one side than on the other.

Inasmuch as every space may be of higher temperature on the right side or on the left side, in turn, it is evident that anomalies the most varied may be met with, and what is of importance is, that these anomalies may persist in some cases for hours.

We have thus far occupied ourselves with qualitative results only; we will now consider the thermometric values of the differences of temperature between symmetrically situated spaces of the two sides of the anterior region.

Investigations of the kind we have now to deal with are more difficult and less satisfactory than those we have been considering. Leaving out the fact that thermo-electric measurements are, at the best, only approximately accurate when applied thermometrically, the greater time consumed in quantitative experiments renders them more liable to be affected by errors due to alterations of temperature of the parts examined than are the briefer qualitative observations. In fact, the head does not usually maintain an even temperature long enough to enable one to accurately compare the whole twenty-seven spaces at a single sitting. It is only by a great number of observations that we can arrive at results at all satisfactory.

Table 2 gives the results of 100 observations on each pair of symmetrically situated spaces of the two sides of the anterior region.

From this table we learn that the mean difference of temperature is pretty nearly the same for both sides of the head:—thus the mean difference of temperature in the eighteen spaces which are, in the majority of cases, warmer on the right side than on the left side is $0.255°$ C. ($0.459°$ F.) ; while the mean difference of temperature in the nine spaces which are warmer on the left side than on the right side, is $0.241°$ C. ($0.433°$ F.). The greatest difference of temperature noted is in the 3rd district, 3rd tier, left side, namely, $0.461°$ C. ($0.829°$ F.). The smallest differences noted are in the 1st district, 4th and 5th tiers, and in the 2nd district, 4th tier, all on the right side,—the differences being each $0.076°$ C. ($0.136°$ F.). The extreme range of difference of temperature is, therefore, $0.385°$ C. ($0.693°$ F.). The mean difference of temperature of all the observations taken together irrespective of sides, is $0.247°$ C. ($0.444°$ F.).

TABLE 2.

Comparison in degrees Centigrade and Fahrenheit of symmetrically situated spaces of the two sides of the anterior region. "Left" denotes left side of the head, and "Right" denotes right side of the head. The figures are placed in the space belonging to that side which has the higher temperature of the two compared: thus the figures 0·138° C.—0·248° F., in the left hand lowest space of the table, signify, that,in the 1st district of the 1st tier, the temperature is, on an average, higher on the left side than on the right side, by the above amounts. The results are the mean of 100 observations on each pair of spaces.

	1st District		2nd District		3rd District		4th District		5th District	
	Left.	Right.	Left.	Right.	Left.	Right.	Left.	Right.	Left.	Right.
6th Tier		0·238°C. 0·428°F.		0·304°C. 0·547°F.						
5th Tier		0·076°C. 0·136°F.		0·192°C. 0·345°F.		0·423°C. 0·761°F.				
4th Tier		0·076°C. 0·136°F.		0·076°C. 0·136°F.		0·384°C. 0·691°F.		0·304°C. 0·547°F.		
3rd Tier	0·204°C. 0·367°F.		0·304°C. 0·547°F.		0·461°C. 0·829°F.	0·304°C. 0·547°F.		0·304°C. 0·547°F.		0·315°C. 0·567°F.
2nd Tier	0·161°C. 0·289°F.		0·384°C. 0·691°F.		0·23°C. 0·414°F.			0·147°C. 0·264°F.		0·306°C. 0·55°F.
1st Tier	0·138°C. 0·248°F.		0·176°C. 0·316°F.		0·114°C. 0·205°F.			0·304°C. 0·547°F.		0·304°C. 0·547°F.

Examination of Anterior Region. 41

If we take the average difference of temperature in each tier, on both sides of the head taken together, we have the following results:

1st Tier.	2nd Tier.	3rd Tier.	4th Tier.	5th Tier.	6th Tier.
0·207°C.	0·215°C.	0·315°C.	0·215°C.	0·259°C.	0·321°C.
(0·372°F.)	(0·377°F.)	(0·567°F.)	(0·377°F.)	(0·466°F.)	(0·577°F.)

Looking at the above figures, we see that the 6th tier has the highest and the 1st tier the lowest mean. As these figures represent the average degree of departure from neutrality of each tier, we would be led to expect that those tiers,—the 3rd and 4th, in which neutrality most frequently occurs,* would show a lower average of difference of temperature than the tiers which are more commonly of higher temperature on one side or the other. Let us see how far our figures will bear out this hypothesis.

Taking first the 3rd tier, we find that so far from showing a lower average than all the rest excepting the 4th tier, it, on the contrary, has the second highest average. With regard to the 4th tier, it is, to be sure, one of the three lowest in average, but then it is equal to the 2nd tier, and superior to the 1st tier. Referring to Tables 1 and 2, we will take from the former the percentages of times of occurrence of neutrality in each of the eleven spaces in which it is most common, and will take from Table 2, and place beside these percentages, the degrees of difference of temperature noted in each of the eleven spaces specified.

	Percentages of times of occurrence of neutrality.	Degree of difference of temperature.
	1st Tier.	
3rd District	16	0·114°C. (0·205°F.).
	2nd Tier.	
3rd District	12	0·23°C. (0·414°F.).
4th „	22	0·147°C. (0·264°F.).
	3rd Tier.	
1st District	22	0·204°C. (0·367°F.).
2nd „	24	0·304°C. (0·547°F.).
3rd „	10	0·461°C. (0·829°F.).
4th „	12	0·304°C. (0·547°F.).

* Vide p. 36.

42   *Temperature of the Head.*

		Percentages of times of occurrence of neutrality.	Degree of difference of temperature.
		4th Tier.	
1st	District	13	0·076°C. (0·136°F.).
2nd	,,	24	0·076°C. (0·136°F.).
3rd	,,	30	0·304°C. (0·547°F.).
4th	,,	12	0·304°C. (0·547°F.).

Examining the above figures we fail to find any definite relation between the frequency of occurrence of neutrality in spaces and the degree of difference of temperature found in these spaces. It is true, that the 3rd district, 3rd tier, shows the smallest percentage of times of occurrence of neutrality and the greatest difference of temperature; but other spaces yield contradictory results:—thus, the highest percentage noted,—30,—is found in a space,—3rd district 4th tier,—which possesses with three others, —4th district, 4th tier, and 2nd and 4th districts, 3rd tier,—the second greatest difference of temperature observed, namely, 0·304° C. (0·547° F.); and, moreover, these three spaces for the same differences of temperature do not all show equal percentages of times of occurrence of neutrality; two,—4th district, 3rd and 4th tiers,—have, indeed, the same percentages, but the third shows twice the percentage of the other two; and another space,—2nd district, 4th tier,—which has the same percentage as the above third-mentioned space, exhibits a difference of temperature only one quarter of that of the latter, namely, 0·076° C. (0·136° F.), the smallest difference of temperature noted, which it has in common with an adjoining space, 1st district, 4th tier,—which latter has but a little more than half the percentage of neutrality of the former.

If we take the average difference of temperature of the eleven spaces enumerated above, and the average difference of temperature of the other sixteen spaces, we find the former figure to be 0·229° C. (0·412° F.), and the latter figure to be 0·260° C. (0·468° F.).

Again, if we take, on the one hand, all those cases in which the percentage of superiority of temperature (right or left, irrespectively) is above 60; and, on the other hand, all those cases in which the percentage of superiority of temperature on a side is 60 or below that amount, we have for the first class of cases,—nine-

Examination of Anterior Region. 43

teen in number,—an average difference of temperature of 0·219° C. (0·394° F.) ; and for the second class of cases,—eight in number, —an average difference of temperature 0·182° C. (0·327° F.).

Lastly, if we take the average difference of temperature of the six spaces which show the highest percentages of cases of superiority of temperature on a side (1st and 2nd districts of 1st and 2nd tiers ; 5th district 2nd tier ; and 3rd district 3rd tier), and the average difference of temperature of the five spaces which show the lowest percentages of cases of superiority of temperature on a side (3rd districts 1st and 2nd tiers ; and 1st, 2nd, and 3rd districts 4th tier), we find the first of these averages to be 0·245° C. (0·441° F.), and the second to be 0·160° C. (0·288° F.).

These last three analyses seem to indicate the existence of some kind of relation between the frequency of occurrence of superiority of tempearture on a side, and the thermometric degree of this superiority. Thus all three analyses give the highest average difference of temperature to those cases in which the average percentage of times of occurrence of superiority of temperature is greatest. The following figures show the percentages of times of occurrence of superiority of temperature on a side, and the corresponding differences of temperature, in each of the above three analyses :—

	Average percentages of times of occurrence of superiority of temperature on a side.	Average difference of temperature.
	Analysis 1.	
In eleven spaces	57·363	0·229°C. (0·412°F.).
In sixteen „	69·875	0·260°C. (0·468°F.).
	Analysis 2.	
In eight spaces	53·375	0·182°C. (0·327°F.).
In nineteen „	69·579	0·219°C. (0·394°F.).
	Analysis 3.	
In five spaces	49·400	0·160°C. (0·288°F.).
In six „	75·333	0·245°C. (0·441°F.).

But in spite of the above results, the discrepancies existing in individual spaces are so numerous and so marked, that it can hardly be conceded that any constant and decided relation exists between the two kinds of values under consideration. To point out some of these individual contradictions ;—the 2nd and 5th districts of the 2nd tier have precisely the same percentages of numbers of times of occurrence of superiority of temperature on a side, and yet the first named of the two spaces shows a difference of temperature more than twice that of the second. Again, the 2nd districts of the 1st and 2nd tiers have within one per cent. the same percentages of superiority of temperature, and yet one shows a difference of temperature double that of the other. Also, the 1st and 3rd districts of the 5th tier, with the same percentages of superiority of temperature, have, respectively, 0·076° C. (0·136° F.), and 0·384° C. (0·69° F.), as differences of temperature, the latter being thus five times greater than the former. Other contradictions may be readily found by comparing Tables 1 and 2.

If we compare the spaces in districts instead of in tiers, we have the following averages of difference of temperature for each district, opposite to which are placed the corresponding percentages of times of occurrence of superiority of temperature on both sides, taken together.

	Average percentages of times of occurrence of superiority of temperature on a side.	Average difference of temperature.
1st District	66·000	0·148°C. (0·266°F.).
2nd ,,	67·000	0·239°C. (0·43°F.).
3rd ,,	58·333	0·319°C. (0·574°F.).
4th ,,	63·800	0·272°C. (0·489°F.).
5th ,,	73·000	0·269°C. (0·484°F.).

Taken thus in districts, the average difference of temperature increases from the 1st district outward to the 3rd district, where it attains its maximum; thence it diminishes in the 4th and 5th districts, but does not in either of these districts fall back to its former points in the 1st and 2nd districts.

The figures do not indicate any relation between the frequency of occurrence of superiority of temperature in a district and the thermometric degree of this superiority. On the contrary, the two lowest percentages of occurrence of superiority of temperature are associated with the two greatest differences of temperature, —3rd and 4th districts.

CHAPTER II.

EXAMINATION OF THE ANTERIOR REGION IN SPACES ON ONE AND THE SAME SIDE.

WE will next consider the relative temperatures of the different spaces on one and the same side of the head. We will first compare the different spaces by tiers, and then by districts.

TABLE 3.

Results of 50 observations on the comparative temperature of each pair of spaces situated in the same district of two adjoining tiers on one and the same side of the anterior region. "Left" and "Right" signify, respectively, left and right sides of the head. The figures denote the number of times, in a total of 50, in which each space was superior in temperature to the space in the same district of the adjoining tier on the same side of the head; thus in 50 comparisons of the 1st and 2nd tiers in the 1st district, on the left side of the head, the temperature was higher in the 1st tier 34 times and in 2nd tier 16 times.

Tiers compared.		1st District.		2nd District.		3rd District.		4th District.		5th District.	
		Left.	Right.	Left.	Right.	Left.	Right.	Left.	Right.	Left.	Right.
1st & 2nd.		28	29	27	29	27	29				
		22	21	23	21	23	21				
2nd & 3rd.		11	12	11	14	17	15	14	18		
		39	38	39	36	33	35	36	32		
3rd & 4th.		20	15	14	15	13	13	10	11	10	12
		30	35	36	35	37	37	40	39	40	38
4th & 5th.		23	22	22	24	24	21	22	23	20	21
		27	28	28	26	26	29	28	27	30	29
5th & 6th.		16	14	35	36	33	35	33	35	39	36
		34	36	15	14	17	15	17	15	11	14

Temperature of the Head.

(*a*) Comparison of spaces situated in the same district of two adjoining tiers (Table 3).

Analyzing the above table we obtain the following results:—

1st. The whole of the 2nd tier is, in the majority of cases of higher temperature than the first tier, on both sides of the head; with the exception that, on both sides, the 1st district is generally of higher temperature in the 1st tier than in the 2nd tier.

2nd. The whole of the 2nd tier is, in the majority of cases, of higher temperature than the 3rd tier, on both sides of the head.

3rd. The whole of the 3rd tier is, in the majority of cases, of higher temperature than the 4th tier, on both sides of the head.

4th. The whole of the 4th tier is, in the majority of cases, of higher temperature than the 5th tier, on both sides of the head.

5th. The whole of the 6th tier is, in the majority of cases, of higher temperature than the 5th tier, on both sides of the head.

It will be observed that there are no cases of equality of temperature in Table 3. Such cases are not, however, unfrequent; but as their places and times of occurrence present no regularity, happening, as they may, in every set of spaces compared, and in very dissimilar ratios of frequency at different times, they have been omitted from the table.

The following are the average percentages of numbers of times of occurrence of superiority of temperature of a given tier over the tier adjoining, on one and the same side of the anterior region:—

Percentages of times of occurrence of superiority of temperature.

	Right side.	Left side.
1st district, 1st tier, superior to 1st district, 2nd tier	72·0	68·000
Remainder of 2nd tier, superior to 1st tier.	71·0	70·000
2nd tier, superior to 3rd tier	55·6	55·600
3rd tier, superior to 4th tier	73·6	73·200
4th tier, superior to 5th tier	70·5	73·500
6th tier, superior to 5th tier	58·0	54·666

TABLE 4.

Comparison in degrees Centigrade and Fahrenheit of each space of each tier with the space in the same district of the tier immediately adjoining on one and the same side of the anterior region. "Left" and "Right" signify respectively left and right sides of the head. The figures are placed in the tier, having the higher temperature of the two compared: thus, the figures 0·04° C.—0·072° F., in the left-hand lowest space of the table, denote that, on the left side of the head, in the 1st district, the temperature is higher in the 1st tier than in the 2nd by the above amount. The results are the mean of 50 observations on each pair of spaces.

	Tiers compared.						
1st & 2nd.	2nd & 3rd.	3rd & 4th.	4th & 5th.	5th & 6th.			
0·04°C. / 0·072°F.	0·11°C. / 0·198°F.	0·31°C. / 0·558°F.	0·2°C. / 0·36°F.	0·2°C. / 0·36°F.	Left.	1st District.	
0·03°C. / 0·054°F.	0·13°C. / 0·234°F.	0·28°C. / 0·504°F.	0·2°C. / 0·36°F.	0·18°C. / 0·324°F.	Right.		
0·17°C. / 0·306°F.	0·08°C. / 0·144°F.	0·35°C. / 0·63°F.	0·33°C. / 0·594°F.	0·32°C. / 0·576°F.	Left.	2nd District.	
0·26°C. / 0·468°F.	0·13°C. / 0·234°F.	0·27°C. / 0·486°F.	0·34°C. / 0·612°F.	0·29°C. / 0·522°F.	Right.		
0·15°C. / 0·27°F.	0·08°C. / 0·144°F.	0·29°C. / 0·522°F.	0·35°C. / 0·63°F.	0·3°C. / 0·54°F.	Left.	3rd District.	
0·23°C. / 0·414°F.	0·09°C. / 0·162°F.	0·26°C. / 0·468°F.	0·36°C. / 0·648°F.	0·31°C. / 0·558°F.	Right.		
0·13°C. / 0·234°F.	0·05°C. / 0·09°F.	0·3°C. / 0·54°F.	0·33°C. / 0·594°F.		Left.	4th District.	
0·1°C. / 0·18°F.	0·03°C. / 0·054°F.	0·31°C. / 0·558°F.	0·35°C. / 0·63°F.		Right.		
0·21°C. / 0·378°F.	0·02°C. / 0·036°F.	0·37°C. / 0·666°F.			Left.	5th District.	
0·03°C. / 0·054°F.	0·02°C. / 0·036°F.	0·36°C. / 0·648°F.			Right.		

Temperature of the Head.

Table 4 gives the thermometric values of the differences of temperature between corresponding spaces of adjoining tiers :— from it we deduce the following results :—

Average thermometric values of differences of temperature between adjoining tiers of one and the same side of the anterior region.

                                Right side.   Left side.

1st district, 1st tier, higher in temperature than 1st district, 2nd tier by . 0·03°C.(0·054°F.)—0·04°C.(0·072°F.).
Remainder of 2nd tier, higher in temperature than remainder of 1st tier by . . 0·155°C.(0·279°F.)—0·165°C (0·297°F.).
2nd tier higher in temperature than 3rd tier by 0·08°C.(0·144°F.)—0·068°C.(0·122°F.).
3rd tier higher in temperature than 4th tier by 0·296°C.(0·532°F.)—0·324°C.(0·583°F.).
4th tier higher in temperature than 5th tier by 0·312°C.(0·561°F.)—0·302°C.(0·543°F.).
6th tier higher in temperature than 5th tier by 0·260°C.(0·468°F.)—0·273°C.(0·491°F.).

The preceding figures show that the 2nd tier has the highest temperature ; that the 3rd tier comes next in order ; then the 1st tier ; after that the 4th ; then the 6th ; and last the 5th ;—and this order holds good for both sides of the head.

If we suppose the average absolute temperature of the 2nd tier, right side, to be 34·5° C. (94·1° F.) ; and that of the same tier, left side, to be 34·608° C. (94·294° F.),* the following will represent the temperature of each tier on the two sides. The first district, 1st tier of each side is omitted from the calculation :—

                  Right side.   Left side.

1st tier—34·345°C. (93·821°F.) . 34·443°C. (93·997°F.).
2nd ,, 34·500°C. (94·1°F.) . 34·608°C. (94·294°F.).
3rd ,, 34·420°C. (93·956°F.) . 34·540°C. (94·172°F.).
4th ,, 34·124°C. (93·424°F.) . 34·216°C. (93·589°F.).
5th ,, 33·812°C. (92·863°F.) . 33·914°C. (93·046°F.).
6th ,, 34·072°C. (93·331°F.) . 34·187°C. (93·537°F.).

* The additional amount 0·108° C. on the left side represents the average superiority of temperature of the left side of the anterior region over the right side in the 2nd tier, as obtained from Table 2.

Examination of Anterior Region. 49

If we take the percentage of times of occurrence of superiority of temperature of one tier over another and the thermometric values of the differences of temperature between the tiers, on both sides of the head taken together, we have the following figures:—

Percentages of times of occurrence of superiority of temperature of one tier over another, on one and the same side of the head. Mean results of both sides taken together.	Mean difference of temperature between adjoining tiers, on one and the same side of the head. Mean results of both sides taken together.

1st district, 1st tier, superior to
1st district, 2nd tier . 70·000 . . 0·035°C. (0·063°F.).
Remainder of 2nd tier, superior
to remainder of 1st tier . 70·500 . . 0·160°C. (0·288°F.).
2nd tier, superior to 3rd tier 55·600 . . 0·074°C. (0·133°F.).
3rd tier, superior to 4th tier 73·400 . . 0·310°C. (0·558°F.).
4th tier, superior to 5th tier 72·000 . . 0·307°C. (0·552°F.).
6th tier, superior to 5th tier 56·333 . 0·266°C. (0·478°F.).

Looking at the above figures, we see that there is no constant relation between the percentage of times in which a given tier is of higher temperature than another tier, and the thermometric value of the difference of temperature between the two tiers. Thus, although the two highest percentages of times of occurrence of superiority of temperature coincide with the two greatest thermometric values (comparison of 3rd tier with 4th tier, and of 4th tier with 5th tier), yet the two lowest percentages coincide with very different thermometric values,—one of these values being more than three times that of the other (comparison of 2nd tier with 3rd tier, and of 6th tier with 5th tier),—the percentages being nearly the same. By comparing Tables 3 and 4, we can find many individual contradictions of the existence of the relation referred to. For example, the comparison of the 4th and 5th tiers, in the 4th district, right side, gives a percentage of superiority to the 4th tier of 64, and a thermometric difference of 0·35° C. (0·63° F.); while the comparison of the 1st and 2nd tiers in the 5th district, right side, gives a percentage of superiority to the 2nd tier of 72, and a thermometric difference of only 0·03° C. (0·054° F.) or a little more than one twelfth of the former difference of temperature.

The average difference of temperature for all the individual spaces, in the comparison by tiers, is, on the right side, 0·207° C. (0·372° F.); and, on the left side, 0·213° C. (0·383° F.). The average difference of temperature for the two sides taken together is, therefore, 0·210° C. (0·378° F.).

We pass now to the comparison by districts of spaces on one and the same side of the anterior region.

(b) Comparison of spaces situated in two adjoining districts in the same tier, on one and the same side of the head.

TABLE 5.

Results of 50 observations on the comparative temperature of each pair of spaces situated in adjoining districts of the same tier on one and the same side of the anterior region. "Left" and "Right" signify, respectively, left and right sides of the head. The figures denote the number of times in which each space was superior in temperature to the space in the adjoining district of the same tier; thus in 50 comparisons of the 1st and 2nd districts in the 1st tier, on the left side, the temperature was higher in the 1st district 30 times, and in the 2nd district 20 times.

Districts compared.

		1st & 2nd.		2nd & 3rd.		3rd & 4th.		4th & 5th.	
6th Tier	Right	18	32	20	30				
	Left	21	29	22	28				
5th Tier	Right	30	20	23	27	19	31		
	Left	30	20	22	28	21	29		
4th Tier	Right	16	34	17	33	16	34	31	19
	Left	13	37	6	44	15	35	36	14
3rd Tier	Right	16	34	16	34	15	35	39	11
	Left	17	33	8	42	17	33	20	30
2nd Tier	Right	13	37	10	40	16	34	33	17
	Left	15	35	15	35	14	36	15	35
1st Tier	Right	28	22	14	36	20	30	22	28
	Left	30	20	18	32	22	28	35	15

The table shows, in the first place, that there is not the same regularity in the distribution of temperature by districts that there is in the distribution by tiers. Analyzing the table, we obtain the following results:—

1st. The 1st district is of higher temperature, in the majority of cases, than the 2nd district in the 1st and 5th tiers, on both sides of the head.

2nd. The 2nd district is of higher temperature than the 1st district, in the majority of cases, in the 2nd, 3rd, 4th, and 6th tiers, on both sides of the head.

3rd. The 3rd district is of higher temperature than the 2nd district, in the majority of cases, in every tier, on both sides of the head.

4th. The 4th district is of higher temperature than the 3rd district, in the majority of cases, on both sides of the head, in every tier.

5th. The 4th district is of higher temperature than the 5th district, in the majority of cases, in the first tier left side; in the 2nd and 3rd tiers, right side; and in the 4th tier on both sides of the head.

6th. The 5th district is of higher temperature than the 4th district, in the majority of cases, in the first tier right side; and in the 2nd and 3rd tiers left side.*

The following are the mean percentages of times of occurrence of superiority of temperature of a given district over the district adjoining.

Percentages of times of occurrence of superiority of temperature.

	Right side.	Left side.
2nd district superior to 1st district	59·666	58·000
3rd district superior to 2nd district	66·666	69·666
4th district superior to 3rd district	65·600	64·400
4th district superior to 5th district	62·500	53·000

* What has been said of the occurrence of equality of temperature in the comparison by tiers, applies also to the comparison by districts; hence the omission from the table of cases of neutrality.

Table 6.

Comparison in degrees Centigrade and Fahrenheit of each space of each district with the space in the adjoining district of the same tier, on one and the same side of the anterior region. "Left" and "Right" signify, respectively, left and right sides of the head. The figures are placed in the district having the higher temperature of the two compared; thus, the figures 0·19° C. (0·342° F.) in the left hand lowest space of the table, denote, that, on the left side of the head, in the 1st tier, the temperature is higher in the 1st district than in the 2nd by these amounts. The results are the mean of 50 observations on each pair of spaces

Districts compared.

		1st and 2nd.	2nd and 3rd.	3rd and 4th.	4th and 5th.
6th Tier	Right		0·01°C. / 0·018°F.	0·07°C. / 0·126°F.		
	Left		0·01°C. / 0·018°F.	0·04°C. / 0·072°F.		
5th Tier	Right	0·1°C. / 0·18°F.		0·05°C. / 0·09°F.	0·09°C. / 0·162°F.
	Left	0·12°C. / 0·216°F.		0·06°C. / 0·108°F.	0·05°C. / 0·09°F.
4th Tier	Right	0·04°C. / 0·072°F.		0·07°C. / 0·126°F.	0·08°C. / 0·144°F.	0·07°C. / 0·126°F.
	Left	0·01°C. / 0·018°F.		0·08°C. / 0·144°F.	0·03°C. / 0·054°F.	0·03°C. / 0·054°F.
3rd Tier	Right	0·03°C. / 0·054°F.		0·06°C. / 0·108°F.	0·13°C. / 0·234°F.	0·02°C. / 0·036°F.
	Left	0·05°C. / 0·09°F.		0·02°C. / 0·036°F.	0·04°C. / 0·072F.	0·04°C. / 0·072°F.
2nd Tier	Right	0·03°C. / 0·054°F.		0·02°C. / 0·036°F.	0·07°C. / 0·126°F.	0·03°C. / 0·054F.
	Left	0·02°C. / 0·036°F.		0·02°C. / 0·036°F.	0·01°C. / 0·018°F.	0·01°C. / 0·018°F.
1st Tier	Right	0·26°C. / 0·468°F.		0·05°C. / 0·09°F.	0·2°C. / 0·36°F.	0·04 C / 0·072°F.
	Left	0·19°C. / 0·342°F.		0·04°C. / 0·072°F.	0·03°C. / 0·054°F.	0·07°C. / 0·126°F.

Examination of Anterior Region. 53

Table 6 gives the thermometric values of the differences of temperature between corresponding spaces of adjoining districts. We obtain from it the following averages.

Average thermometric values of differences of temperature between adjoining districts of one and the same side of the anterior region.

          Right side.   Left side.

1st district higher in temperature than 2nd district by . 0·042° C. (0·075° F.)—0·037° C. (0·066°F.).
3rd district higher in temperature than 2nd district by . 0·053° C. (0·095° F.)—0·043° C. (0·077° F.).
4th district higher in temperature than 3rd district by . 0·114° C. (0·205° F.)—0·032° C. (0·057° F.).
4th district higher in temperature than 5th district by . 0·02° C. (0·036° F.)—0·012° C. (0·021° F.).

From the above figures we learn that the 4th district has the highest temperature; that next in order comes the 5th district; then the 3rd district; then the 1st district; and lastly, the 2nd district; this order holding good of both sides of the head.

Taking the average temperature of the 4th district of the right side as 34·5° C. (94·1° F.), and that of the same district of the left side as 34·228° C. (93·61° F.),* we obtain the following temperatures for each side of the head, taken by districts:

1st district—34·375° C. (93·875° F.)—34·190° C. (93·542° F.).
2nd „ 34·333° C. (93·8° F.) —34·153° C. (93·476° F.).
3rd „ 34·386° C. (93·895° F.)—34·196° C. (93·553° F.).
4th „ 34·500° C. (94·1° F.) —34·228° C. (93·610° F.).
5th „ 34·480° C. (94·064° F.) —34·216° C. (93·589° F.).

Comparing the average percentages of times of occurrence of superiority of temperature of one district over another with the corresponding average thermometric values of difference of tem-

* The 0·272° C. difference between the right and left sides is the average difference between the 4th districts of the two sides of the head, according to Table 2.

perature, on both sides of the head considered together, we have the following figures:—

	Percentages of times of occurrence of superiority of temperature of one district over another on one and the same side of the head. Mean results of both sides taken together.	Mean difference of temperature between adjoining districts on one and the same side of the head. Mean results of both sides taken together.
1st district superior to 2nd district	59·00	0·167° C. (0·3° F.).
2nd district superior to 1st district	67·75	0·025° C. (0·045° F.).
3rd district superior to 2nd district	68·166	0·048° C. (0·086° F.).
4th district superior to 3rd district	64·8	0·073° C. (0·131° F.).
4th district superior to 5th district	69·6	0·044° C. (0·079° F.).
5th district superior to 4th district	62·0	0·03° C. (0·054° F.).

Here again we fail to find a definite relation between the frequency of occurrence of superiority of temperature and the thermometric value of this superiority; and a comparison of Tables 5 and 6 shows that the individual spaces also fail to indicate any constant relation between the two classes of values.

The average difference of temperature for all the individual spaces in the comparison by districts is, on the right side, 0·063° C. (0·113° F.); and, on the left side, 0·046° C. (0·083° F.). The average difference of temperature for the two sides taken together is, therefore, 0·055° C. (0·099° F.).

Lastly, if we take the three principal classes of experiments, which have thus far engaged our attention, namely, those on the comparative temperature of the two sides of the head; those on the comparative temperature of different tiers on one and the same side of the head; and those on the comparative temperature of different districts of one and the same side of the head, we have the following values:—

Comparison of the two sides of the head.

Average percentage of times of occurrence of superiority of temperature of either side of the head over the other . . 64·777—

Average difference of temperature

0·247° C. (0·444° F.).

Comparison of adjoining tiers of one and the same side.

Average percentage of times of occurrence of superiority of temperature of one tier over another on both sides taken together . . . . 66·045—

Average difference of temperature.

0·21° C. (0·378° F.).

Comparison of adjoining districts of one and the same side.

Average percentage of times of occurrence of superiority of temperature of one district over another on both sides taken together . . . . . 66·119—

Average difference of temperature.

0·055° C. (0·099° F.).

According to the above figures, superiority of temperature of one side of the head over the other occurs a little less frequently than superiority of temperature of one tier or district over another on one and the same side of the head.* With regard to degree of difference of temperature the greatest average is found where the two sides of the head are compared; the next greatest average (very nearly equal to the first), where tiers on one and the same side are compared; and the smallest average (decidedly less than the other two), where districts on one and the same side are compared.

* The two sides of the head are rarely equal in temperature, but the balance of superiority shifts so frequently from one side to the other that the percentage of superiority for either side, in a given number of observations, is comparatively small.

CHAPTER III.

SUBDIVISIONS AND MEASUREMENTS OF THE MIDDLE REGION.
—EXAMINATION OF THE MIDDLE REGION IN SYMMETRI-
CALLY SITUATED SPACES OF THE TWO SIDES.

Subdivisions and measurements of the middle region.

THE boundaries of the middle region have been given on p. 29. This region is divided on each side into seven tiers by six equidistant lines drawn parallel to the longitudinal median line. The tiers are numbered 1 to 7 from below upward. There are five districts in each lateral half, formed by four equidistant lines drawn parallel to the anterior and posterior boundaries. The districts are numbered 1 to 5 from the anterior boundary backward. The following are the measurements of the standard head :—

Height of each lateral half of the region, measured on the anterior boundary line, 159·5 mm. (6·279 inches). Height of each lateral half, measured on the posterior boundary line, 190 mm. (7·48 inches).* Breadth of region on the median line 83 mm. (3·26 inches). The height of each lateral half on the anterior boundary being 159·5 mm. (6·279 inches), each tier will measure vertically on this line 22·78 mm. (0·89 inch). The breadth of the region on the median line being 83 mm. (3·26 inches), each district will measure horizontally on this line 16·6 mm. (0·65 inch). The ear cuts off horizontally 32·5 mm. (1·27 inch) ; commencing at 4·3 mm. (0·169 inch) from the anterior boundary of the 3rd district 1st tier, it includes the remainder of the 3rd district, the whole of the 4th district, and 3·6 mm. (0·14 inch) of the 5th district. Vertically, the attachments of the ear to the temporal bone extend 3 mm. into the 2nd tier in the 3rd and 4th districts. The upper boundary of the 4th tier is 96 mm. (3·779 inches) distant, vertically, from the upper border of the external auditory meatus. The posterior boundary of the 3rd district in

---

* These two measurements of height are the halves of the distances over the top of the head, taken on the anterior and posterior boundaries, respectively.

the 1st tier, passes through the centre of the external auditory meatus.

As it was impossible on account of the hair to measure off the middle region with the chalked string used in the anterior region, callipers were employed. Where the hair was thick the piles were made sufficiently small to pass between the hairs down to the skin (p. 21, Part I).

We will proceed to examine the middle region in the same manner as that adopted in the anterior region. We have thirty-four spaces on a side to examine.

Comparison of symmetrically situated spaces of the two sides of the head.

Table 7.

Results of 100 observations on the comparative temperature of each pair of symmetrically-situated spaces of the two sides of the middle region. "R." signifies right side, "L." left side, and "N." equality of the two sides. The figures prefixed to "R.," "L.," and "N.," denote the number of times in a total of 100 in which the temperature was higher on the right side, on the left side, or was equal on the two sides.

	1st district.	2nd district.	3rd district.	4th district.	5th district.
7th Tier.	30 L. 65 R. 5 N.	54 L. 40 R. 6 N.	70 L. 30 R.	72 L. 28 R.	60 L. 40 R.
6th Tier.	36 L. 64 R.	59 L. 29 R. 12 N.	75 L. 25 R.	73 L. 23 R. 4 N.	70 L. 30 R.
5th Tier.	35 L. 65 R.	63 L. 29 R. 8 N.	77 L. 17 R. 6 N.	65 L. 30 R. 5 N.	60 L. 35 R. 5 N.
4th Tier.	38 L. 62 R.	49 L. 42 R. 9 N.	70 L. 26 R. 4 N.	52 L. 41 R. 7 N.	50 L. 40 R. 10 N.
3rd Tier.	35 L. 65 R.	40 L. 60 R.	35 L. 65 R.	34 L. 66 R.	33 L. 67 R.
2nd Tier.	35 L. 65 R.	28 L. 72 R.	33 L. 67 R.	32 L. 68 R.	28 L. 72 R.
1st Tier.	60 L. 30 R. 10 N.	35 L. 54 R. 11 N.	30 L. 65 R. 5 N.	Ear.	40 L. 60 R.

Temperature of the Head.

Examining the table we see that in the middle region, as in the anterior region, every space may be of higher temperature on either the right side or the left side, in turn.

Omitting the instances of equality of temperature we have for each district of each tier the following proportions in numbers of times of superiority of temperature on the right and left sides respectively:

1st Tier.

1st District—2·00 times to 1 in favour of left side.*
2nd ,, 1·54 ,, ,, right side.
3rd ,, 2·17 ,, ,, ,,
4th ,, ear.
5th ,, 1·5 ,, ,, ,,

2nd Tier.

1st District—1·86 times to 1 in favour of right side.
2nd ,, 2·57 ,, ,,
3rd ,, 2·03 ,, ,,
4th ,, 2·12 ,, ,,
5th ,, 2·57 ,, ,,

3rd Tier.

1st District—1·86 times to 1 in favour of right side.
2nd ,, 1·50 ,, ,,
3rd ,, 1·86 ,, ,,
4th ,, 1·94 ,, ,,
5th ,, 2·03 ,, ,,

4th Tier.

1st District—1·63 times to 1 in favour of right side.
2nd ,, 1·16 ,, ,, left side.
3rd ,, 2·69 ,, ,, ,,
4th ,, 1·27 ,, ,, ,,
5th ,, 1·25 ,, ,, ,,

5th Tier.

1st District—1·86 times to 1 in favour of right side.
2nd ,, 2·17 ,, ,, left side.
3rd ,, 4·53 ,, ,, ,,
4th ,, 2·17 ,, ,, ,,
5th ,, 1·71 ,, ,, ,,

* M. Broca's "frontal" thermometer would seem to have been placed either in this space or in the 5th district 1st tier, anterior region,—or again, possibly, partly in each space.

6th Tier.

1st District—1·78 times to 1 in favour of right side.
2nd   „   2·03   „   „   left side.
3rd   „   3·00   „   „   „
4th   „   3·17   „   „   „
5th   „   2·33   „   „   „

7th Tier.

1st District—2·17 times to 1 in favour of right side.
2nd   „   1·35   „   „   left side.
3rd   „   2·33   „   „   „
4th   „   2·57   „   „   „
5th   „   1·50   „   „   „

The above shows us that, of the thirty-four spaces on a side compared, one half are of higher temperature on each side.

Taking the total number of observations in which either the right or left side was the warmer, namely, 3293, we find that the right side had the higher temperature in 1637 instances, and that the left side had the higher temperature in 1656 instances. The percentages of times of occurrence of superiority of temperature are, therefore, for the right side 49·711, and for the left side 50·289. The mean percentage in favour of the right side in the seventeen spaces which, in the majority of cases, are of higher temperature on this side is 65·634, and the corresponding percentage in favour of the left side is 66·852.

The spaces in which neutrality occurs, and the proportionate numbers of times of its occurrence in each space, are as follows:

1st Tier.

1st District—1 in 10·00 times.
2nd   „   1 „ 9·09   „
3rd   „   1 „ 20·00   „

4th Tier.

2nd District—1 in 11·11 times.
3rd   „   1 „ 25·00   „
4th   „   1 „ 14·28   „
5th   „   1 „ 10·00   „

5th Tier.
2nd District—1 in 12·500 times.
3rd   „   1 „ 16·666   „
4th   „   1 „ 20·000   „
5th   „   1 „ 20·000   „

6th Tier.
2nd District—1 in 8·333 times.
4th   „   1 „ 25·000   „

7th Tier.
1st District—1 in 20·000 times.
2nd   „   1 „ 16·666   „

Thus equality of temperature of the two sides is found in fifteen spaces, or in 44·117 per cent. of the whole number of spaces. Of the total number of observations included in the table,—3400—107, or 3·147 per cent., show equality of temperature. In the fifteen spaces to which equality of temperature is limited it exists in 7·133 per cent. of all the observations.

In the total number of observations we have the following percentages of times of occurrence of superiority of temperature on the right and left sides, and of equality of temperature on the two sides:

In favour of right side—48·147 per cent.
    „   left   „   48·706   „
    „   equality   3·147   „

In the fifteen spaces in which right and left superiority and equality of temperature, are all found, the percentage of each condition is as follows:

In favour of right side—37·733 per cent.
    „   left   „   55·134   „
    „   equality   7·133   „

These latter figures show a decrease in the percentage for the right side, and an increase in the percentages for the left side and for equality.

If now we compare the results thus far given with the corresponding ones obtained in the anterior region, we observe the following differences:

1st. The distribution of superiority of temperature by numbers of spaces is equal on the two sides, in the middle region, whereas, in the anterior region, two thirds of the spaces are in favour of the right side.

2nd. The times of occurrence of superiority of temperature on the right and left sides, respectively, are more nearly equal in the middle region than in the anterior region. Thus, in the former region the percentages are—right side, 49·711 ; left side, 50·289 ; while in the latter region the percentages are—right side 54·153 ; left side 45·847.

3rd. Equality of temperature has both a more limited area and occurs in a smaller percentage of times in the middle region than in the anterior region.

Taking the percentages of right and left superiority of temperature, and of equality of temperature for each entire tier, we have the following values :

Percentages for Tiers.

	Right side.	Left side.	Neutral.
1st Tier	52·25	41·25	6·5
2nd ,,	68·80	31·20	0
3rd ,,	64·60	35·40	0
4th ,,	42·20	51·80	6·0
5th ,,	35·20	60·00	4·8
6th ,,	34·20	62·60	3·2
7th ,,	40·60	57·20	2·2

The 2nd tier shows the highest percentage for the right side ; the 6th tier the highest percentage for the left side ; and the 1st tier the highest percentage for equality.

Percentages for Districts.

	Right side.	Left side.	Neutral.
1st District	59·428	38·428	2·144
2nd ,,	46·571	46·857	6·572
3rd ,,	42·143	55·714	2·143
4th ,,	42·666	54·666	2·668
5th ,,	49·143	48·714	2·143

The 1st district shows the highest percentage for the right side ; the 3rd district the highest percentage for the left side ; and the 2nd district the highest percentage for neutrality.

With regard to individual spaces the highest percentages are as follows :

In favour of Right Side.

2nd District of 2nd Tier—72 per cent.
3rd   ,,   2nd ,, 67 ,,
4th   ,,   2nd ,, 68 ,,
5th   ,,   2nd ,, 72 ,,
5th   ,,   3rd ,, 67 ,,

In favour of Left Side.

3rd District of 5th Tier—77 per cent.
3rd   ,,   6th ,, 75 ,,
4th   ,,   6th ,, 73 ,,
4th   ,,   7th ,, 72 ,,

In favour of Neutrality.

1st District of 1st Tier—10 per cent.
2nd   ,,   1st ,, 11 ,,
2nd   ,,   4th ,, 9 ,,
5th   ,,   4th ,, 10 ,,
2nd   ,,   6th ,, 12 ,,

To sum up the principal points of the comparative distribution of superiority of temperature on the two sides of the middle region thus far given we find, that, commencing with the 1st district, 1st tier, the temperature in this space is higher on the left side; that in the remaining three districts of the 1st tier, and in the whole of the 2nd and 3rd tiers, the balance of higher temperature is in favour of the right side; that, the 1st district in the 4th, 5th, 6th, and 7th tiers is also of higher temperature on the right side; that, in the 2nd, 3rd, 4th, and 5th districts of the 4th, 5th, 6th, and 7th tiers the higher temperature is found on the left side (Diagram 5).

Unlike the anterior region the middle region presents no marked anomalies of distribution of temperature.

DIAGRAM 5.

Distribution of comparative superiority of temperature on the two sides of the middle region.

The shaded spaces are those which, of symmetrically situated spaces on the two sides, have the higher temperature. "Right" and "Left" signify respectively right and left sides of the head.

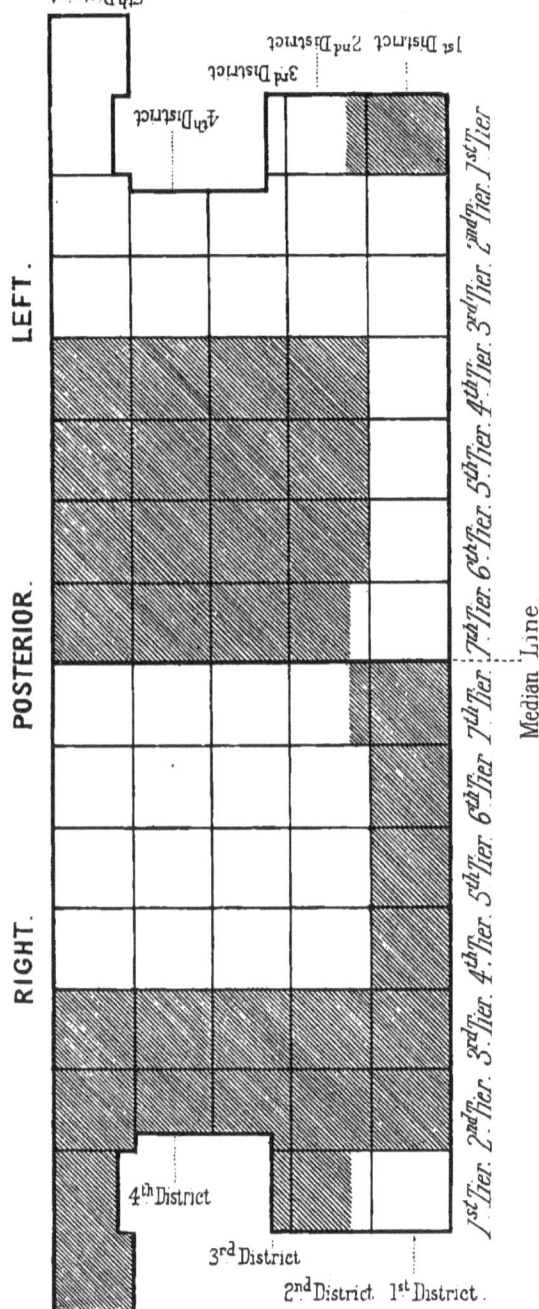

Examination of Middle Region.

TABLE 8.

Comparison in degrees Centigrade and Fahrenheit of symmetrically-situated spaces of the two sides of the middle region. "Left" and "Right" denote, respectively, left and right sides of the head. The figures are placed in the space belonging to that side which has the higher temperature of the two compared : thus, the figures 0·033°C.—0·0594°F. in the left hand lowest space of the table signify that in the 1st district of the 1st tier, the temperature is, on an average, higher on the left side than on the right side, by the above amount. The results are the mean of 100 observations on each pair of spaces.

	1st District		2nd District		3rd District		4th District		5th District	
	Left.	Right.	Left.	Right.	Left.	Right.	Left.	Right.	Left.	Right.
7th Tier		0·049°C. 0·088°F.	0·066°C. 0·1188°F.							
6th Tier		0·082°C. 0·147°F.	0·082°C. 0·147°F.		0·214°C. 0·385°F.		0·156°C. 0·28°F.		0·115°C. 0·207°F.	
5th Tier		0·115°C. 0·197°F.	0·082°C. 0·147°F.		0·181°C. 0·325°F.		0·181°C. 0·325°F.		0·156°C. 0·28°F.	
4th Tier		0·066°C. 0·1188°F.	0·041°C. 0·073°F.		0·264°C. 0·475°F.		0·165°C. 0·297°F.		0·016°C. 0·0288°F.	
3rd Tier		0·049°C. 0·088°F.		0·066°C. 0·1188°F.	0·033°C. 0·0594°F.	0·082°C. 0·147°F.		0·033°C. 0·0594°F.	0·024°C. 0·0432°F.	
2nd Tier		0·033°C. 0·0594°F.		0·082°C. 0·147°F.		0·033°C. 0·0594°F.	0·066°C. 0·1188°F.	0·016°C. 0·0288°F.		0·016°C. 0·0288°F.
1st Tier	0·033°C. 0·0594°F.			0·09°C. 0·162°F.		0·057°C. 0·102°F.	Ear			0·198°C. 0·356°F.

Analyzing the above table, we find that, unlike the anterior region, the middle region shows a decided inequality in the mean differences of temperature of the two sides:—thus, the mean difference of temperature for the seventeen spaces, which are, on an average, of higher temperature on the right side than on the left, is 0·0589° C. (0·106° F.); while the mean difference of temperature for the seventeen spaces, which are, on an average, of higher temperature on the left side than on the right, is 0·1103° C. (0·198° F.). The greatest difference of temperature noted is in the 3rd district, 5th tier, left side, namely, 0·264° C. (0·475° F.); the smallest differences noted are in the 4th district, 2nd tier, and 5th district, 2nd and 3rd tiers, all right side,—the difference in each case being 0·016° C. (0·028° F.). The extreme range of difference of temperature is, therefore, 0·248° C. (0·446° F.). The mean difference of temperature of all the observations taken together, irrespective of sides, is 0·087° C. (0·152° F.).

Taking, next, the average difference of temperature in each tier, on both sides of the head taken together, we have the following values:—

1st Tier.	2nd Tier.	3rd Tier.	4th Tier.	5th Tier.	6th Tier.	7th Tier.
0·094°C.	0·036°C.	0·049°C.	0·046°C.	0·128°C.	0·136°C.	0·12°C.
(0·169°F.)	(0·064°F.)	(0·088°F.)	(0·082°F.)	(0·23°F.)	(0·244°F.)	(0·216°F).

The 6th tier has the highest, and the 2nd tier the lowest, average.

Adopting the course pursued in the examination of the anterior region, let us see if any relation exists between the frequency of occurrence of neutrality in a tier and the mean difference of temperature exhibited by that tier.

The 1st and 4th tiers show the highest, and the 7th tier the lowest, percentages of neutrality. In the 2nd and 3rd tiers neutrality is absent. Now, the 1st and 4th tiers show (especially the 4th tier) comparatively small differences of temperature; and the 7th tier shows one of the three greatest differences of temperature; but, to offset these results, the 2nd and 3rd tiers, in both of which neutrality is wanting, instead of showing a high average of difference of temperature, show, the one, the lowest, and, the other, one of the three lowest, averages of differences of temperature noted. If we take the percentages of the fifteen

Examination of Middle Region. 65

spaces in which equality of temperature is found, and place opposite to each percentage the average difference of temperature observed for that space, we have the following values for comparison :

	Percentages of times of occurrence of neutrality.	Degree of difference of temperature.
	1st Tier.	
1st District	10	0·033° C. (0·0594° F.).
2nd ,,	11	0·09° C. (0·162° F.).
3rd ,,	5	0·057° C. (0·102° F.).
	4th Tier.	
2nd District	9	0·041° C. (0·073° F.).
3rd ,,	4	0·033° C. (0·594° F.).
4th ,,	7	0·066° C. (0·1188° F.).
5th ,,	10	0·024° C. (0·0432° F.).
	5th Tier.	
2nd District	8	0·082° C. (0·147° F.).
3rd ,,	6	0·264° C. (0·475° F.).
4th ,,	5	0·165° C. (0·297° F.).
5th ,,	5	0·016° C. (0·0288° F.).
	6th Tier.	
2nd District	12	0·082° C. (0·147° F.).
4th ,,	4	0·181° C. (0·325° F.).
	7th Tier.	
1st District	5	0·049° C. (0·088° F.).
2nd ,,	6	0·066° C. (0·1188° F.).

The three greatest differences of temperature (3rd and 4th districts, 5th tier, and 4th district, 6th tier) are here associated with comparatively small percentages of neutrality; and the four largest percentages of neutrality (1st and 2nd districts, 1st tier, 5th district, 4th tier, and 2nd district, 6th tier) are associated with comparatively small differences of temperature. But contradictory results are obtained on further analysis. Thus the 4th and 5th districts of the 5th tier, with equal percentages, have very unequal differences of temperature, that of the former being more than ten times that of the latter. One of the two smallest percentages is found with a low average of difference of temperature (3rd district, 4th tier), this average being, moreover, less than

Temperature of the Head.

half that associated with the greatest percentage of all noted (2nd district, 6th tier). Further, the smallest difference of temperature noted is found with one of the lowest percentages (5th district, 5th tier).

To ascertain if any relation exists between the frequency of occurrence of superiority of temperature on a side and the thermometric value of the difference observed, we will make three analyses after the manner of those prepared in the examination of the anterior region.

Analysis 1.

	Average percentage or times of occurrence of superiority of temperature on a side.	Average difference of temperature.
In fifteen spaces in which neutrality exists	61·066	0·0832° C. (0·1497° F.).
In nineteen spaces in which neutrality is absent	66·578	0·0899° C. (0·1618° F.).

Analysis 2.

In twenty-three spaces in which the percentage of superiority of temperature is above 60	67·956	0·0946° C. (0·1703° F.).
In eleven spaces in which the percentage of superiority of temperature is 60 or below	56·182	0·0724° C. (0·1304° F.).

Analysis 3.

In six spaces having the highest percentages of superiority of temperature	73·5	0·146° C. (0·263° F.).
In six spaces having the lowest percentages of superiority of temperature	53·0	0·0615° C. (0·1107° F.).

All three of the above analyses are in favour of a relation

Examination of Middle Region. 67

between the frequency of occurrence of superiority of temperature on a side and the thermometric value of this superiority, since in each analysis the greater percentage of frequency coincides with the greater thermometric difference. If we take the individual spaces, we find that the highest percentage, 77 (3rd district, 5th tier), coincides with the greatest thermometric value, 0·264° C. (0·475° F.). But further examination shows a number of discrepancies. Thus the third greatest difference of temperature, 0·198° C. (0·365° F.), (5th district, 1st tier) is associated with a percentage of 60, while three spaces which show a difference of temperature less than one twelfth of the above—0·016° C. (0·0288° F.)—have percentages of 66, 67, and 68 respectively (4th and 5th districts, 2nd tier; and 5th district, 3rd tier). The five smallest differences of temperature, averaging 0·0176° C. (0·031° F.), are associated with an average percentage of 64·2 (4th district, 2nd tier; 5th district, 2nd tier; 5th district, 4th tier; 5th district, 3rd tier; and 5th district, 5th tier); while, as Analysis 3 shows, an average difference, three and a half times greater than that just given, is found with an average percentage of only 53. Again, the 3rd district, 2nd tier, and 3rd district, 4th tier, with the same difference of temperature, 0·033° C. (0·0594° F.), have, respectively, as percentages, 67 and 70; while the 4th district, 5th tier, and 5th district, 7th tier, with differences of temperature, 0·165° C. (0·297° F.), and 0·115° C. (0·207° F.) respectively, are associated, the former with a percentage of 65, and the latter with a percentage of 60.

In the next place, comparing the spaces in districts, we have the following averages of difference of temperature for each district, and opposite to these values are placed the average corresponding percentages of times of occurrence of superiority of temperature.

	Average percentage of times of occurrence of superiority of temperature on a side.	Average difference of temperature.
1st District	63·714	0·061° C. (0·109° F.).
2nd ,,	58·714	0·072° C. (0·129° F.).
3rd ,,	69·857	0·123° C. (0·221° F.).
4th ,,	66·000	0·103° C. (0·185° F.).
5th ,,	62·714	0·077° C. (0·138° F.).

68   *Temperature of the Head.*

The 3rd district has the greatest average difference of temperature, the 4th district the second greatest average, and next in order come, successively, the 5th, 2nd, and 1st districts.

This last comparison would seem, in some degree, to favour the existence of the relationship we have just been considering, so far as comparison by districts is concerned. Thus the two highest percentages of times of occurrence of superiority of temperature and the two greatest differences of temperature are found in the same districts (3rd and 4th); but the results in the other three districts are in opposition to this conclusion. The lowest percentage does not coexist with the smallest difference of temperature (2nd district); and the 1st district, with a larger percentage than the 5th district, shows a smaller difference of temperature.

CHAPTER IV.

EXAMINATION OF THE MIDDLE REGION IN SPACES ON ONE AND THE SAME SIDE.

WE will now examine the relative temperatures of spaces on one and the same side of the head, first by tiers and then by districts.

Comparison of spaces situated in the same district of two adjoining tiers.

TABLE 9.

Results of 50 observations on the comparative temperature of each pair of spaces situated in the same district of two adjoining tiers on one and the same side of the middle region. "Left" and "Right" signify, respectively, left and right sides of the head. The figures denote the number of times, in a total of 50, in which each space was superior in temperature to the space in the same district of the adjoining tier, on the same side of the head; thus, in 50 comparisons of the 1st and 2nd tiers in the 1st district, on the

Examination of Middle Region. 69

left side, the temperature was higher in the 1st tier thirty-six times, and in the 2nd tier fourteen times.

	1st District		2nd District		3rd District		4th District		5th District	
	Left.	Right.	Left.	Right.	Left.	Right.	Left.	Right.	Left.	Right.
6th & 7th	20	22	23	24	23	24	24	22	20	17
	30	28	27	26	27	26	26	28	30	33
5th & 6th	18	16	19	17	17	16	27	30	32	34
	32	34	31	33	33	34	23	20	18	16
4th & 5th	20	19	29	28	29	23	18	16	12	16
	30	31	21	22	21	27	32	34	38	34
3rd & 4th	12	13	15	14	15	16	13	15	17	18
	38	37	35	36	35	34	37	35	33	32
2nd & 3rd	15	16	14	14	17	16	13	16	17	14
	35	34	36	36	33	34	37	34	33	36
1st & 2nd	14	17	19	16	18	15	Ear.		30	33
	36	33	31	34	32	35			20	17

On analysis of Table 9 we obtain the following results:

1st. The whole of the 1st tier is, in the majority of cases, of higher temperature than the 2nd tier, on both sides of the head; with the exception, that, on both sides of the head, the 5th district is generally of higher temperature in the 2nd tier than in the 1st tier.

2nd. The whole of the 2nd tier is, in the majority of cases, of higher temperature than the 3rd tier, on both sides of the head.

3rd. The whole of the 3rd tier is, in the majority of cases, of higher temperature than the 4th tier, on both sides of the head.

4th. The 4th tier is of higher temperature than the 5th tier, in the majority of cases, in the 1st, 4th, and 5th districts, on both sides of the head, and in the 3rd district on the right side; in the 2nd district, on both sides, and in the 3rd district, on the left side, the temperature is higher in the 5th tier than in the 4th tier.

5th. The 5th tier is of higher temperature than the 6th tier, in the majority of cases, in the 1st, 2nd, and 3rd districts, on both sides of the head; in the 4th and 5th districts, on both sides of the head, the 6th tier is, in the majority of cases, of higher temperature than the 5th tier.

6th. The 6th tier is of higher temperature than the 7th tier, in the majority of cases, in every district of both sides of the head.*

The following are the average percentages of times of occurrence of superiority of temperature of a given tier over the tier adjoining, on one and the same side of the head, in the middle region :

Percentages of times of occurrence of superiority of temperature.

	Right side.	Left side.
5th district, 2nd tier, superior to 5th district, 1st tier	66·0	60·0
Remainder of 1st tier, superior to remainder of 2nd tier	68·0	66·0
2nd tier, superior to 3rd tier	69·6	69·6
3rd tier, superior to 4th tier	69·6	71·2
4th tier, superior to 5th tier	59·2	56·8
5th tier, superior to 6th tier	54·8	54·8
6th tier, superior to 7th tier	56·4	52·0

Table 10 gives the thermometric values of the differences of temperature between the spaces of adjoining tiers. Analyzing the table, we obtain the following results :

Average thermometric differences of temperature between adjoining tiers of one and the same side of the middle region.

	Right side.	Left side.
5th district, 2nd tier, higher in temperature than 5th district, 1st tier, by	0·082°C.(0·147°F.)	0·264°C.(0·475°F.)
Remainder of 1st tier higher in temperature than remainder of 2nd tier by	0·109°C.(0·196°F.)	0·12°C (0·216°F.)

* What has been said with reference to equality of temperature in comparisons by tiers and districts, in the examination of the anterior region, is also applicable to the middle and posterior regions.

TABLE 10.

Comparison in degrees, Centigrade and Fahrenheit, of each space of each tier, with the space in the same district of the tier immediately adjoining, on one and the same side of the middle region. The figures are placed in the tier having the higher temperature of the two compared. "Left" and "Right" signify, respectively, left and right sides of the head:—thus, the figures 0·082° C.—0·147° F.,—in the left hand lowest space of the table, denote, that, on the left side of the head, in the 1st district, the temperature is higher in the first tier than in the 2nd tier by the above amount. The results are the mean of fifty observations on each pair of spaces.

Tiers compared.	1st District. Left.	1st District. Right.	2nd District. Left.	2nd District. Right.	3rd District. Left.	3rd District. Right.	4th District. Left.	4th District. Right.	5th District. Left.	5th District. Right.
6th and 7th (7th tier)										
6th and 7th (6th tier)	0·082°C. 0·147°F.	0·115°C. 0·207°F.	0·082°C. 0·147°F.	0·066°C. 0·1188°F.	0·074°C. 0·133°F.	0·107°C. 0·192°F.	0·14°C. 0·252°F.	0·115°C. 0·207°F.	0·156°C. 0·28°F.	0·115°C. 0·207°F.
5th and 6th (6th tier)							0·049°C. 0·088°F.	0·033°C. 0·0594°F.	0·338°C. 0·608°F.	0·198°C. 0·356°F.
5th and 6th (5th tier)	0·074°C. 0·133°F.	0·107°C. 0·192°F.	0·288°C. 0·518°F.	0·288°C. 0·518°F.	0·371°C. 0·667°F.	0·288°C. 0·518°F.				
4th and 5th (5th tier)			0·115°C. 0·207°F.	0·074°C. 0·133°F.	0·156°C. 0·28°F.					
4th and 5th (4th tier)	0·066°C. 0·1188°F.	0·016°C. 0·0288°F.				0·074°C. 0·133°F.	0·231°C. 0·415°F.	0·33°C. 0·594°F.	0·47°C. 0·846°F.	0·462°C. 0·831°F.
3rd and 4th (4th tier)										
3rd and 4th (3rd tier)	0·247°C. 0·444°F.	0·231°C. 0·415°F.	0·198°C. 0·356°F.	0·305°C. 0·549°F.	0·033°C. 0·0594°F.	0·148°C. 0·266°F.	0·033°C. 0·0594°F.	0·132°C. 0·237°F.	0·024°C. 0·0432°F.	0·066°C. 0·1188°F.
2nd and 3rd (3rd tier)										
2nd and 3rd (2nd tier)	0·354°C. 0·637°F.	0·338°C. 0·608°F.	0·272°C. 0·489°F.	0·288°C. 0·518°F.	0·099°C. 0·178°F.	0·049°C. 0·088°F.	0·033°C. 0·0594°F.	0·016°C. 0·0288°F.	0·033°C. 0·0594°F.	0·033°C. 0·0594°F.
1st and 2nd (2nd tier)							Ear.	Ear.	0·264°C. 0·475°F.	0·082°C. 0·147°F.
1st and 2nd (1st tier)	0·082°C. 0·147°F.	0·016°C. 0·0288°F.	0·016°C. 0·0288°F.	0·024°C. 0·0432°F.	0·264°C. 0·475°F.	0·288°C. 0·518°F.				

Temperature of the Head.

                                Right side.   Left side.

2nd tier higher in temperature than 3rd tier by 0·144°C.(0·259°F.)—0·158°C.(0·284°F.).
3rd tier higher in temperature than 4th tier by 0·176°C.(0·316°F.)—0·107°C.(0·192°F.).
4th tier higher in temperature than 5th tier by 0·161°C.(0·289°F.)—0·099°C.(0·178°F.).
5th tier higher in temperature than 6th tier by 0·09°C.(0·162°F.)—0·069°C.(0·124°F.).
6th tier higher in temperature than 7th tier by 0·103°C.(0·185°F.)—0·106°C.(0·19°F.).

The 1st tier has the highest average absolute temperature; then come, in regular order, the 2nd, 3rd, 4th, 5th, 6th, and 7th tiers, this order holding good for both sides of the head.

Taking the average absolute temperature of the 1st tier right side, as 33·81° C. (92·858° F.); and that of the same tier left side, as 33·732° C. (92·717° F.), we have the following average absolute temperatures for each tier of each side:

            Right side.   Left side.
1st tier—33·81°C. (92·858°F.) . 33·732°C. (92·717°F.).
2nd „ 33·682°C. (92·627°F.) . 33·646°C. (92·562°F.).
3rd „ 33·537°C. (92·366°F.) . 33·488°C. (92·278°F.).
4th „ 33·360°C. (92·048°F.) . 33·381°C. (92·085°F.).
5th „ 33·199°C. (91·758°F.) . 33·282°C. (91·907°F.).
6th „ 33·109°C. (91·596°F.) . 33·212°C. (91·781°F.).
7th „ 33·005°C. (91·409°F.) . 33·105°C. (91·589°F.).

Comparing the percentages of times of occurrence of superiority of temperature of one tier over another and the thermometric values of the differences of temperature between the tiers, we have the following figures:

	Percentages of times of occurrence of superiority of temperature of one tier over another, on one and the same side of the head. Mean results of both sides taken together.	Mean differences of temperature between adjoining tiers, on one and the same side of the head. Mean results of both sides taken together.
5th district, 2nd tier, superior to 5th district, 1st tier . . .	630	0·173° C. (0·32° F.).

Examination of Middle Region.

	Percentages of times, &c.	Mean differences, &c.
Remainder of 1st tier superior to remainder of 2nd tier	67·0	0·114° C. (0·205° F.).
2nd tier superior to 3rd tier	69·6	0·151° C. (0·271° F.).
3rd tier superior to 4th tier	70·4	0·141° C. (0·253° F.).
4th tier superior to 5th tier	58·0	0·13° C. (0·234° F.).
5th tier superior to 6th tier	54·8	0·0795° C. (0·143° F.).
6th tier superior to 7th tier	54·2	0·1045° C. (0·188° F.).

The above figures fail to show any definite relation between the frequency with which one tier is superior in temperature to another, and the theromometric difference between the tiers. The greatest difference of temperature (comparison of 5th districts 1st and 2nd tiers) is associated with a percentage of frequency fourth in value on the list. A percentage of frequency of 67 is found with a difference of temperature of 0·114° C. (0·205° F.), while a percentage of frequency of 58 is associated with a thermometric value of 0·13° C. (0·234° F.).

The average difference of temperature for all the individual spaces, in the comparison by tiers, is, on the right side, 0·151° C. (0·271° F.) ; and, on the left side, 0·160° C. (0·288° F.). The average difference of temperature for the two sides taken together is, therefore, 0·1555° C. (0·279° F.).

Comparison of spaces situated in two adjoining districts in the same tier on one and the same side of the head.

TABLE II.

Results of 50 observations on the comparative temperature of each pair of spaces situated in adjoining districts of the same tier, on one and the same side of the middle region. " Left " and " Right " signify, respectively, left and right sides of

Temperature of the Head.

the head. The figures denote the number of times in a total of 50 in which each space was superior in temperature to the space in the adjoining district of the same tier, on the same side of the head:—thus, in 50 comparisons of the 1st and 2nd districts in the 1st tier, on the left side, the temperature was higher in the 1st district twenty-eight times, and in the 2nd district twenty-two times.

Districts compared.

		1st & 2nd.		2nd & 3rd.		3rd & 4th.		4th & 5th.	
7th Tier	Right	33	17	34	16	34	16	37	13
	Left	32	18	35	15	37	13	39	11
6th Tier	Right	31	19	37	13	34	16	34	16
	Left	33	17	33	17	33	17	32	18
5th Tier	Right	34	16	36	14	32	18	36	14
	Left	18	32	34	16	33	17	33	17
4th Tier	Right	33	17	35	15	34	16	34	16
	Left	32	18	33	17	32	18	31	19
3rd Tier	Right	30	20	36	14	36	14	32	18
	Left	29	21	37	13	37	13	34	16
2nd Tier	Right	28	22	32	18	35	15	33	17
	Left	29	21	35	15	31	19	36	14
1st Tier	Right	27	23	31	19	Ear.			
	Left	28	22	31	19				

Analyzing the above table we obtain the following results:

1st. The 1st district is, in the majority of cases, of higher temperature than the 2nd district in every tier, on both sides of the head, with the exception of the 5th tier, on the left side, where the temperature is higher in the 2nd district than in the 1st district.

2nd. The 2nd district is, in the majority of cases, of higher temperature than the 3rd district, in every tier, on both sides of the head.

Examination of Middle Region. 75

3rd. The 3rd district is, in the majority of cases, of higher temperature than the 4th district, in every tier, on both sides of the head.

4th. The 4th district is, in the majority of cases, of higher temperature than the 5th district, in every tier, on both sides of the head.

The following are the mean percentages of times of occurrence of superiority of temperature of a given district over the district adjoining :

Percentages of times of occurrence of superiority of temperature.

	Right side.	Left side.
1st district superior to 2nd district	61·714	57·428
2nd district superior to 3rd district	68·857	68·000
3rd district superior to 4th district	68·333	67·666
4th district superior to 5th district	68·666	68·333

Table 12 gives the thermometric values of the differences of temperature between adjoining districts. Analyzing the table we obtain the following results :

Average thermometric values of differences of temperature between adjoining districts of one and the same side of the middle region.

	Right side.	Left side.
1st district higher in temperature than 2nd district by	0·127° C. (0·228° F.)	0·0707° C. (0·127°F.)
2nd district higher in temperature than 3rd district by	0·216° C. (0·388° F.)	0·147° C. (0·264° F.)
3rd district higher in temperature than 4th district by	0·115° C. (0·207° F.)	0·124° C. (0·223° F.)
4th district higher in temperature than 5th district by	0·0793° C. (0·142° F.)	0·119° C. (0·214° F.)

The 1st district has the highest average absolute temperature, and then follow, in regular order, the 2nd, 3rd, 4th, and 5th districts, this order holding good for both sides of the head.

Temperature of the Head.

TABLE 12.

Comparison in degrees Centigrade and Fahrenheit of each space of each district with the space in the adjoining district of the same tier, on one and the same side of the middle region. "Left" and "Right" signify, respectively, left and right sides of the head. The figures are placed in the district having the higher temperature of the two compared; thus, the figures 0·231° C. (0·415° F.) in the left hand lowest space of the table, denote that, on the left side of the head, in the 1st tier the temperature is higher in the 1st district than in the 2nd by the above amounts. The results are the mean of 50 observations on each pair of spaces.

Districts compared.

		1st and 2nd.	2nd and 3rd.	3rd and 4th.	4th and 5th.
7th Tier	Right	0·181°C. / 0·325°F.	0·239°C. / 0·43°F.	0·024°C. / 0·0432°F.	0·016°C. / 0·0288°F
	Left	0·066°C. / 0·1188°F.	0·09°C. / 0·162°F.	0·082°C. / 0·147°F.	0·057°C. / 0·102°F.
6th Tier	Right	0·231°C. / 0·415°F.	0·198°C. / 0·356°F.	0·016 C. / 0·0288°F.	0·016°C. / 0·0288°F.
	Left	0·066°C. / 0·1188°F.	0·099°C. / 0·178°F.	0·016°C. / 0·0288°F.	0·041°C. / 0·0738°F.
5th Tier	Right	0·049°C. / 0·088°F.	0·198°C. / 0·356°F.	0·338°C. / 0·608°F.	0·181°C. / 0·325°F.
	Left	0·148°C. / 0·266°F.	0·016°C. / 0·0288°F.	0·437°C. / 0·786°F.	0·33°C. / 0·594°F.
4th Tier	Right	0·14°C. / 0·252°F.	0·049°C. / 0·088°F.	0·082°C. / 0·147°F.	0·049°C. / 0·088°F.
	Left	0·033°C. / 0·0594°F.	0·057°C. / 0·102°F.	0·049°C. / 0·088°F.	0·09°C. / 0·162°F.
3rd Tier	Right	0·066°C. / 0·1188°F.	0·206°C. / 0·37°F.	0·099°C. / 0·178°F.	0·115°C. / 0·207°F.
	Left	0·082°C. / 0·147°F.	0·222°C. / 0·399°F.	0·049°C. / 0·088°F.	0·099°C. / 0·178°F.
2nd Tier	Right	0·115°C. / 0·207°F.	0·445°C. / 0·801°F.	0·132°C. / 0·237°F.	0·099°C. / 0·178°F.
	Left	0·165°C. / 0·297°F.	0·396°C. / 0·712°F.	0·115°C. / 0 207°F.	0·099°C. / 0·178°F.
1st Tier	Right	0·107°C. / 0·192°F.	0·181°C. / 0·325°F.	Ear.	
	Left	0·231°C. / 0·415°F.	0·148°C. / 0·266°F.		

Examination of Middle Region.

Taking the average absolute temperature of the 1st district, right side, as 33·689° C. (92·64° F.), and that of the same district, left side, as 33·638° C. (92·548° F.), we have the following average absolute temperatures for each district of each side :

	Right side.	Left side.
1st district—	33·689° C. (92·64° F.)	—33·638° C. (92·548° F.).
2nd „	33·562° C. (92·411° F.)	—33·5673° C. (92·421° F.).
3rd „	33·346° C. (92·022° F.)	—33·42° C. (92·156° F.).
4th „	33·231° C. (91·815° F.)	—33·296° C. (91·932° F.).
5th „	33·151° C. (91·671° F.)	—33·177° C. (91·718° F.).

Comparing the average percentages of times of occurrence of superiority of temperature of one district over another with the average thermometric values of differences of temperature between the districts, on both sides of the head taken together, we have the following figures :

	Percentages of times of occurrence of superiority of temperature of one district over another, on one and the same side of the head. Mean results of both sides taken together.	Mean difference of temperature between adjoining districts on one and the same side of the head. Mean results of both sides taken together.
1st district superior to 2nd district	59·571	0·0988° C. (0·177° F.).
2nd district superior to 3rd district	68·428	0·181° C. (0·325° F.).
3rd district superior to 4th district	67·999	0·119° C. (0·214° F.).
4th district superior to 5th district	68·499	0·099° C. (0·178° F.).

The above results fail to show any definite relation between frequency of occurrence of superiority of temperature and degree of difference of temperature.

The average difference of temperature for all the individual spaces, in the comparison by districts, is, on the right side, 0·137° C. (0·246° F.) ; and on the left side, 0·126° C. (0·226° F.). The average difference of temperature for the two sides taken together is, therefore, 0·131° C. (0·235° F.).

78   *Temperature of the Head.*

Taking the three principal classes of observations, as was done with the anterior region, we have the following values :

Comparison of the two sides of the head.

Average percentage of times of occurrence of superiority of temperature of either side of the head over the other . . 64·147—
Average difference of temperature
0·087° C. (0·156° F.).

Comparison of adjoining tiers of one and the same side.

Average percentage of times of occurrence of superiority of temperature of one tier over another on both sides taken together . . . . . . 64·724—
Average difference of temperature.
0·1555° C. (0·279° F.).

Comparison of adjoining districts of one and the same side.

Average percentage of times of occurrence of superiority of temperature of one district over another on both sides taken together . . . . . 65·346—
Average difference of temperature.
0·131° C. (0·235° F.).

The above figures show that in the middle region, as in the anterior region, superiority of temperature of one side of the head over the other occurs a little less frequently than superiority of temperature of one tier or district over another, on one and the same side of the head. The comparison by districts here, as in the case of the anterior region, shows the greatest percentage.

With regard to the quantitative results, the greatest difference is found in the comparison by tiers, the second greatest in the comparison by districts, and the smallest in the comparison of the two sides of the head.

CHAPTER V.

SUBDIVISIONS AND MEASUREMENTS OF THE POSTERIOR REGION.—
EXAMINATION OF THE POSTERIOR REGION IN SYMMETRI-
CALLY SITUATED SPACES OF THE TWO SIDES.

THE boundaries of the posterior region have been given on page 29. This region is divided on each side into six tiers by five equidistant, horizontal lines, drawn from the median line to the lateral boundaries. The tiers are numbered 1 to 6 from below upward. Each lateral half is further divided into five districts by four equidistant lines drawn parallel to the median line. The districts are numbered 1 to 5 from the median line outward. The following are the measurements of the standard head :

Height of region measured on median line 128 mm. (4·03 inches). Breadth of whole region measured on a horizontal line passing through the occipital protuberance 210 mm. (8·26 inches). On the median line each tier measures vertically 21·33 mm. (0·83 inch). On the horizontal line passing through the occipital protuberance each district measures 21 mm. horizontally. As in the case of the anterior region, the 5th tier has only four districts, and the 6th tier lacks both the 4th and 5th districts. We have twenty-seven spaces a on side to examine in the posterior region. We will follow the order of examination adopted in the anterior and middle regions.

Comparison of symmetrically situated spaces of the two sides of the head.

TABLE 13.

Results of 100 observations on the comparative temperature of each pair of symmetrically situated spaces of the two sides of the posterior region. " R." signifies right side ; " L.," left side ; and " N.," equality of the two sides. The figures prefixed to " L.," " R.," and " N.," denote the number of times in a total of 100 in which the temperature was higher on the left side, on the right side, or was equal on the two sides.

	1st district.	2nd district.	3rd district.	4th district.	5th district
6th Tier.	60 L. 40 R.	70 L. 30 R.	69 L. 31 R.		
5th Tier.	68 L. 32 R.	67 L. 33 R.	72 L. 28 R.	66 L. 34 R.	
4th Tier.	70 L. 30 R.	72 L. 21 R. 7 N.	65 L. 25 R. 10 N.	73 L. 20 R. 7 N.	65 L. 31 R. 4 N.
3rd Tier.	65 L. 35 R.	59 L. 35 R. 6 N.	30 L. 65 R. 5 N.	35 L. 60 R. 5 N.	31 L. 69 R.
2nd Tier.	65 L. 23 R. 12 N.	36 L. 56 R. 8 N.	29 L. 71 R.	38 L. 62 R.	30 L. 70 R.
1st Tier.	63 L. 30 R. 7 N.	30 L. 61 R. 9 N.	28 L. 72 R.	36 L. 64 R.	37 L. 63 R.

As in the anterior and middle regions, every space in the posterior region may be of higher temperature on the right side or on the left side, in turn.

Leaving out for the present the cases of equality of temperature, we have for each district of each tier the following proportions in numbers of times of occurrence of superiority of temperature on the right and left sides respectively.

1st Tier.

1st District—2·1 times to 1 in favour of left side.
2nd   ,,   2·03   ,,   ,,   right side.
3rd   ,,   2·57   ,,   ,,   ,,
4th   ,,   1·78   ,,   ,,   ,,
5th   ,,   1·70   ,,   ,,   ,,

Examination of Posterior Region.

2nd Tier.
1st District—2·82 times to 1 in favour of left side.
2nd „   1·55   „   „   right side.
3rd „   2·45   „   „   „
4th „   1·63   „   „   „
5th „   2·33   „   „   „

3rd Tier.
1st District—1·86 times to 1 in favour of left side.
2nd „   1·68   „   „   „
3rd „   2·17   „   „   right side.
4th „   1·71   „   „   „
5th „   2·23   „   „   „

4th Tier.
1st District—2·33 times to 1 in favour of left side.
2nd „   3·43   „   „   „
3rd „   2·6   „   „   „
4th „   3·65   „   „   „
5th „   2·09   „   „   „

5th Tier.
1st District—2·12 times to 1 in favour of left side.
2nd „   2·03   „   „   „
3rd „   2·57   „   „   „
4th „   1·94   „   „   „

6th Tier.
1st District—1·50 times to 1 in favour of left side.
2nd „   2·33   „   „   „
3rd „   2·23   „   „   „

According to the above figures, eleven spaces are of higher temperature on the right side and sixteen spaces are of higher temperature on the left side.

Taking the total number of observations, 2700, and deducting 80 cases of equality of temperature, the following is the apportionment of the remaining 2620 cases :

In favour of right side - - - 1191
 „   „   left   „   - - - 1429

Hence the percentages of times of occurrence of superiority of temperature for the right and left sides, respectively, are 45·458 and 54·542. The mean percentage for the eleven spaces, which are, in the majority of cases, of higher temperature on the right side, is 66·449; and the mean percentage for the sixteen spaces in which the higher temperature is more commonly found on the left side is 69·102.

The following is the distribution and the proportionate numbers of times of occurrence of equality of temperature :

1st Tier.

1st District—1 in 14·28 times.
2nd   ,,   1 ,, 11·11   ,,

2nd Tier.

1st District—1 in 8·33 times.
2nd   ,,   1 ,, 12·50   ,,

3rd Tier.

2nd District—1 in 16·66 times.
3rd   ,,   1 ,, 20·00   ,,
4th   ,,   1 ,, 20·00   ,,

4th Tier.

2nd District—1 in 14·28 times.
3rd   ,,   1 ,, 10·00   ,,
4th   ,,   1 ,, 14·28   ,,
5th   ,,   1 ,, 25·00   ,,

From the above, we see that equality of temperature is found in eleven spaces or in 40·74 per cent. of the whole number of spaces.

Taking the total—2700—of all the observations, we have the following percentages of times of occurrence of superiority of temperature on the right and left sides, and of equality of temperature on the two sides :

In favour of right side—44·112 per cent.
      ,,   left   ,,   52·926   ,,
      ,,   equality   2·962   ,,

In the eleven spaces in which right and left superiority and equality of temperature are all found, the percentage of each condition is as follows :

In favour of right side—38·819 per cent.
„ left „ 53·909 „
„ equality 7·272 „

Let us now compare the principal results thus far obtained in the posterior region with the corresponding results found in the anterior and middle regions :

In the first place, the posterior region, like the anterior region, and unlike the middle region, shows inequality in the distribution by numbers of spaces of superiority of temperature, eleven spaces having the higher temperature on the right side and sixteen spaces having the higher temperature on the left side. Unlike the anterior region, however, the balance is in favour of the left side, in the posterior region.

The times of occurrence of superiority of temperature on the right and left sides, respectively, are unequal in the posterior region, as we have seen to be the case in the anterior region. But, whereas the balance in the anterior region is in favour of the right side, in the posterior region the left side predominates.

Equality of temperature has both a more limited distribution, and occurs less frequently in the posterior region than in either of the other regions ; the difference in percentage of times of occurrence of this condition between the middle and posterior regions is, however, insignificant.

If we take the percentages of right and left superiority of temperature and of equality of temperature for each entire tier, we have the following values :

Percentages for Tiers.

	Right side.	Left side.	Neutral.
1st Tier	58·0	38·8	3·2
2nd „	56·4	39·6	4·0
3rd „	52·8	44·0	3·2
4th „	25·4	69·0	5·6
5th „	31·75	68·25	0
6th „	33·666	66·334	0

The 1st tier has the highest percentage for the right side ; the 4th tier the highest percentage for the left side, and also for equality.

Temperature of the Head.

Percentages of Districts.

	Right side.	Left side.	Neutral.
1st District	31·667	65·167	3·166
2nd ,,	39·333	55·667	5·000
3rd ,,	48·666	48·834	2·500
4th ,,	48·000	49·600	2·400
5th ,,	58·250	40·750	1·000

The 5th district shows the highest percentage for the right side; the 1st district the highest percentage for the left side; and the second district the highest percentage for equality.

If we take the individual spaces, the highest percentages are as follows:

In favour of Right Side.

3rd District of 1st Tier—72 per cent.
3rd ,, 2nd ,, 71 ,,
5th ,, 2nd ,, 70 ,,
3rd ,, 3rd ,, 65 ,,
5th ,, 3rd ,, 69 ,,

In favour of Left Side.

1st District of 4th Tier—70 per cent.
2nd ,, 4th ,, 72 ,,
4th ,, 4th ,, 73 ,,
3rd ,, 5th ,, 72 ,,
2nd ,, 6th ,, 70 ,,

In favour of Neutrality.

2nd District of 1st Tier— 9 per cent.
1st ,, 2nd ,, 12 ,,
2nd ,, 2nd ,, 8 ,,
3rd ,, 4th ,, 10 ,,

Summing up the principal points regarding the distribution of superiority of temperature on the two sides of the posterior region, it will be seen, that the whole of the 1st district; the whole of 4th, 5th, and 6th tiers, and the greater part of the 2nd district of the 3rd tier, are of higher temperature on the left side than on the right side. The 2nd, 3rd, 4th, and 5th districts of the 1st and 2nd tiers; the 3rd, 4th, 5th, and part of the 2nd districts, of the 3rd tier, are of higher temperature on the right side than on the left side. [Diagram 6.]

DIAGRAM 6.

Distribution of comparative superiority of temperature on the two sides of the posterior region.

The shaded spaces are those which, of symmetrically situated spaces on the two sides, have the higher temperature. "Right" and "Left" signify respectively right and left sides of the head.

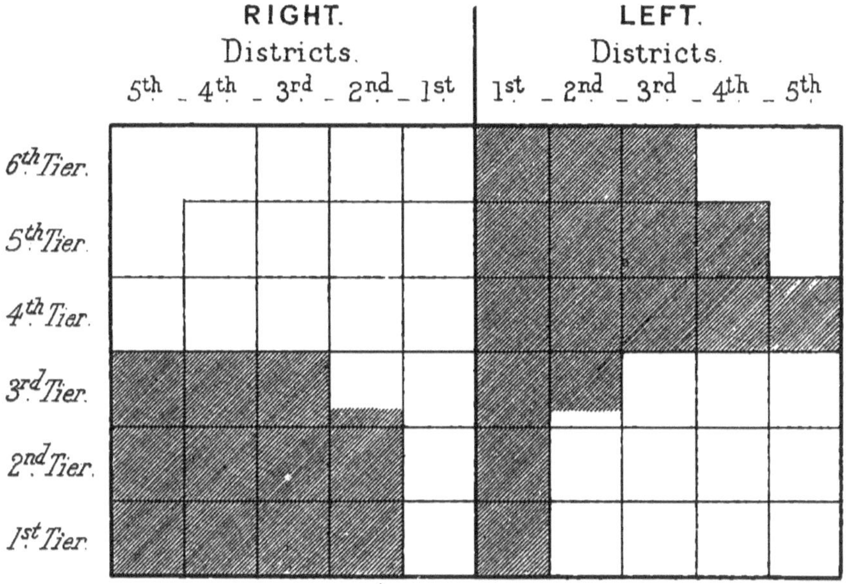

RIGHT. | LEFT.
Districts. | Districts.
5th _ 4th _ 3rd _ 2nd _ 1st | 1st _ 2nd _ 3rd _ 4th _ 5th

6th Tier.
5th Tier.
4th Tier.
3rd Tier.
2nd Tier.
1st Tier.

Examination of Posterior Region. 85

The posterior region presents no marked anomalies of distribution of temperature.

Quantitative comparisons of the two sides of the posterior region.

TABLE 14.

Comparison in degrees Centigrade and Fahrenheit of symmetrically situated spaces of the two sides of the posterior region. " Left " and " Right " denote, respectively, left and right sides of the head. The figures are placed in the space belonging to that side which has the higher temperature of the two compared: thus, the figures 0·043° C.— 0·0774° F.—in the left hand lowest space of the table denote, that, in the 1st district of the 1st tier, the temperature is, on an average, higher on the left side than on the right side, by these amounts. The results are the mean of 100 observations on each pair of spaces.

	1st District.		2nd District.		3rd District.		4th District.		5th District.	
	Left.	Right.	Left.	Right.	Left.	Right.	Left.	Right.	Left.	Right.
6th Tier	0·08°C. 0·144°F.		0·162°C. 0·291°F.		0·097°C. 0·174°F.					
5th Tier	0·038°C. 0·068°F.		0·06°C. 0·108°F.		0·095°C. 0·171°F.		0·021°C. 0·037°F.			
4th Tier	0·021°C. 0·037°F.		0·008°C. 0·014°F.		0·032°C. 0·576°F.		0·06°C. 0·108°F.		0·066°C. 0·1188°F.	
3rd Tier	0·064°C. 0·115°F.		0·113°C. 0·203°F.			0·315°C. 0·567°F.	0·386°C. 0·694°F.			0·228°C. 0·41°F.
2nd Tier	0·097°C. 0·174°F.			0·112°C. 0·201°F.		0·097°C. 0·174°F.	0·241°C. 0·433°F.			0·161°C. 0·289°F.
1st Tier	0·043°C. 0·0774°F.			0·072°C. 0·129°F.		0·055°C. 0·099°F.	0·241°C. 0·433°F.			0·14°C. 0·252°F.

Examining the above table, we perceive, that, like the middle region, and unlike the anterior region, there is, in the posterior region, a marked difference in the values representing the mean differences of temperature in favour of the right and left sides, respectively:—thus, the mean difference of temperature for the eleven spaces which are, in the majority of cases, of higher temperature on the right side, is 0·186° C. (0·334° F.); while the

Temperature of the Head.

mean difference of temperature for the sixteen spaces which are, in the majority of cases, of higher temperature on the left side, is 0·066° C. (0·118° F.). The greatest difference noted is in the 4th district 3rd tier, right side, namely, 0·386° C. (0·694° F.).; the smallest difference noticed is in the 2nd district 4th tier, left side, namely, 0·008° C. (0·0144° F.). The extreme range of difference of temperature is, therefore, 0·378° C. (0·68° F). The mean difference of all the observations taken together irrespective of sides is 0·115° C. (0·207° F.).

The following are the average differences of temperature for each tier, on both sides of the head taken together:

1st Tier.	2nd Tier.	3rd Tier.	4th Tier.	5th Tier.	6th Tier.
0·11°C.	0·141°C.	0·221°C.	0·037°C.	0·053°C.	0·113°C.
(0·198°F.)	(0·253°F.)	(0·397°F.)	(0·066°F.)	(0·095°F.)	(0·203°F.)

The 3rd tier has the highest, and the 4th tier the lowest, average.

We will next see if any relation can be found between the frequency of occurrence of neutrality in a tier, and the mean difference of temperature exhibited by that tier.

The 4th tier has the greatest percentage of neutrality and the least difference of temperature; but the greatest difference of temperature (3rd tier) is associated with the second greatest percentage of neutrality. So little difference exists, however, between the average percentages of times of occurrence of neutrality in the different tiers that it will be necessary to examine the spaces individually. Following, therefore, the course pursued in the anterior and middle regions, we take the respective percentages of equality of temperature in the eleven spaces in which this condition is observed, and compare these values with the corresponding differences of temperature exhibited by the eleven spaces.

	Pecentages of times of occurrence of neutrality.	Degree of difference of temperature.
1st Tier.		
1st District	7	0·043° C. (0·0774° F.).
2nd ,,	9	0·072° C. (0·129° F.).
2nd Tier.		
1st District	12	0·097° C. (0·174° F.).
2nd ,,	8	0·112° C. (0·207° F.).

3rd Tier.

	Percentages of times of occurrence of neutrality.	Degree of difference of temperature.
2nd District	6	0·113° C. (0·203° F.).
3rd ,,	5	0·315° C. (0·567° F.).
4th ,,	5	0·386° C. (0·694° F.).

4th Tier.

2nd District	7	0·008° C. (0·014° F.).
3rd ,,	10	0·032° C. (0·057° F.).
4th ,,	7	0·06° C. (0·108° F.).
5th ,,	4	0·066° C. (0·1188° F.).

In the above comparisons there would appear to be evidence that a high average difference of temperature is associated with a low percentage of frequency of occurrence of neutrality; thus, the three greatest differences of temperature are coincident with three of the four lowest percentages of neutrality (3rd tier). Also, the second highest percentage of neutrality is associated with the second smallest difference of temperature (3rd district, 4th tier). The highest percentage of neutrality (12 per cent.) is not, however, associated with the smallest difference of temperature (0·008° C.). Nor is the lowest percentage (4 per cent.) associated with, by any means, a great difference of temperature (0·066° C.). The highest percentage (12) is associated with a difference of temperature (0·097° C.) more than twice that which accompanies a percentage of only 7 (1st district, 1st tier). Moreover, a percentage of 4 is associated with a difference of 0·066° C., while a percentage of 8 is associated with a difference of 0·112° C.

We will, in the next place, make three analyses, as was done in the examinations of the anterior and middle regions, to ascertain if any relation exists between the frequency of occurrence of superiority of temperature on a side and the thermometric value of the difference observed.

88   *Temperature of the Head.*

Analysis 1.

	Average percentage of times of occurrence of superiority of temperature on a side.	Average difference of temperature.
In eleven spaces in which neutrality exists	64	0·118° C. (0·212° F.).
In sixteen spaces in which neutrality is absent	67·375	0·112° C. (0·201° F.).

Analysis 2.

In thirteen spaces in which the percentage of superiority of temperature is above 65	69·923	0·0848° C. (0·152° F.).
In fourteen spaces in which the percentage of superiority of temperature is 65 or below	62·357	0·143° C. (0·257° F.).

Analysis 3.

In five spaces having the highest percentages of superiority of temperature	72·0	0·063° C. (0·113° F.).
In five spaces having the lowest percentages of superiority of temperature	59·2	0·152° C. (0·273° F.).

Contrary to what was observed in the anterior and middle regions, all three of the above analyses indicate a coincidence of greater percentage of frequency of superiority of temperature with smaller difference of temperature. The contradictory results prove that no fixed relation exists between the two classes of values.

If we compare the spaces in districts we have the following average differences of temperature, opposite to which are placed the corresponding average percentages of times of occurrence of superiority of temperature. Both sides are estimated together as before :

Examination of Posterior Region. 89

	Average percentage of times of occurrence of superiority of temperature on a side.	Average difference of temperature.
1st District	65·166	0·0571° C. (0·102° F.).
2nd „	64·166	0·087° C. (0·156° F.).
3rd „	69·000	0·115° C. (0·207° F.).
4th „	65·000	0·189° C. (0·34° F.).
5th „	66·750	0·149° C. (0·262° F.).

The 4th district has the greatest average difference of temperature; then follow, in the order named, the 5th, 3rd, 2nd, and 1st districts.

CHAPTER VI.

EXAMINATION OF THE POSTERIOR REGION IN SPACES ON ONE AND THE SAME SIDE.—COMPARISON OF THE THREE REGIONS WITH EACH OTHER.—ABSOLUTE TEMPERATURES OF THE THREE REGIONS.

WE next pass to the examination of the relative temperatures of spaces on one and the same side of the head.

Comparison of spaces situated in the same district of two adjoining tiers.

TABLE 15.

Results of 50 observations on the comparative temperature of each pair of spaces situated in the same district of two adjoining tiers on one and the same side of the posterior region. "Left" and "Right" signify, respectively, left and right sides of the head. The figures denote the numbers of times, in a total of 50, in which each space was superior in temperature to the space in the same district of the adjoining tier, on the same side of the head; thus in 50 comparisons of the 1st and 2nd tiers in the 1st district, on the left side, the temperature was higher in the 1st tier 34 times, and in the 2nd tier 16 times.

Tiers compared		1st District Left	1st District Right	2nd District Left	2nd District Right	3rd District Left	3rd District Right	4th District Left	4th District Right	5th District Left	5th District Right
1st & 2nd		21	18	22	23	20	20				
		29	32	28	27	30	30				
2nd & 3rd		22	21	18	21	19	16	20	23		
		28	29	32	29	31	34	30	27		
3rd & 4th		35	31	32	34	29	23	33	31	29	17
		15	19	18	16	21	27	17	19	21	33
4th & 5th		21	21	23	21	20	23	17	21	19	20
		29	29	27	29	30	27	33	29	31	30
5th & 6th		16	13	20	17	20	22	34	33	29	30
		34	37	30	33	30	28	16	17	21	20

Analyzing Table 15, we arrive at the following facts:

1st. The 1st tier is, in the majority of cases, of higher temperature than the 2nd tier, on both sides of the head, in the 1st, 2nd, and 3rd districts. In the 4th and 5th districts, on both sides of the head, the 2nd tier is, in the majority of cases, of higher temperature than the 1st tier.

2nd. The 2nd tier is, in the majority of cases, of higher temperature than the 3rd tier, in every district, on both sides of the head.

3rd. The 4th tier is, in the majority of cases, of higher temperature than the 3rd tier in every district, on both sides of the head, with the exceptions of the 3rd and 5th districts, right side, in both of which spaces the 3rd tier is of higher temperature than the 4th tier, in the majority of cases.

4th. The 4th tier is, in the majority of cases, of higher temperature than the 5th tier in every district, on both sides of the head.

5th. The 5th tier is, in the majority of cases, of higher temperature than the 6th tier in every district, on both sides of the head.

Examination of Posterior Region.

The following are the average percentages of times of occurrence of superiority of temperature of one tier over another, in the posterior region.

Percentages of times of occurrence of superiority of temperature.

	Right side.	Left side.
4th and 5th districts, 2nd tier, superior to 4th and 5th districts, 1st tier	63·000	63·000
Remainder of 1st tier, superior to remainder of 2nd tier	65·333	62·666
2nd tier superior to 3rd tier	57·600	60·000
4th tier superior to 3rd tier	64·000	63·200
3rd tier superior to 4th tier in 3rd and 5th districts	60·000	60·00
4th tier superior to 5th tier	59·500	61·000
5th tier superior to 6th tier	59·333	58·000

Table 16 gives the thermometric values of the differences of temperature between the spaces of adjoining tiers. Analyzing the table we have the following results :

Average thermometric differences of temperature between adjoining tiers on one and the same side of the posterior region.

Right side. Left side.
1st tier higher in temperature than 2nd tier by 0·103°C.(0·185°F.)—0·112°C.(0·201°F.).
2nd tier higher in temperature than 1st tier, in 4th and 5th districts, by 0·0725°C.(0·13°F.)—0·062°C·(0·111°F.).
2nd tier higher in temperature than 3rd tier by 0·157°C.(0·282°F.)—0·204°C.(0·368°F.).

Temperature of the Head.

TABLE 16.

Comparison in degrees Centigrade and Fahrenheit of each space of each tier with the space in the same district of the tier immediately adjoining, on one and the same side of the posterior region. The figures are placed in the tier having the higher temperature of the two compared. "Left" and "Right" signify, respectively, left and right sides of the head; thus, the figures 0·04° C. (0·072° F.), in the left hand lowest space of the table, denote that, on the left side of the head, in the 1st district, the temperature is higher in the 1st tier than in the 2nd tier by these amounts. The results are the mean of 50 observations on each pair of spaces.

District	Side	1st and 2nd	2nd and 3rd	3rd and 4th	4th and 5th	5th and 6th
5th District	Right			0·011°C / 0·198°F	0·017°C / 0·0306°F	0·104°C / 0·187°F
5th District	Left		0·184°C / 0·331°F		0·084°C / 0·151°F	0·083°C / 0·149°F
4th District	Right		0·399°C / 0·718°F	0·033°C / 0·0594°F	0·128°C / 0·23°F	0·041°C / 0·073°F
4th District	Left	0·438°C / 0·788°F	0·479°C / 0·862°F		0·273°C / 0·491°F	0·041°C / 0·073°F
3rd District	Right	0·13°C / 0·234°F	0·372°C / 0·669°F	0·06°C / 0·108°F	0·016°C / 0·0288°F	0·094°C / 0·169°F
3rd District	Left	0·128°C / 0·23°F	0·310°C / 0·558°F	0·288°C / 0·518°F	0·234°C / 0·421°F	0·136°C / 0·244°F
2nd District	Right	0·177°C / 0·318°F	0·255°C / 0·459°F	0·464°C / 0·835°F	0·368°C / 0·662°F	0·122°C / 0·219°F
2nd District	Left	0·075°C / 0·135°F	0·203°C / 0·365°F	0·359°C / 0·646°F	0·144°C / 0·259°F	0·162°C / 0·291°F
1st District	Right	0·099°C / 0·178°F	0·112°C / 0·201°F	0·243°C / 0·437°F	0·256°C / 0·46°F	0·094°C / 0·169°F
1st District	Left	0·057°C / 0·102°F	0·095°C / 0·171°F	0·2°C / 0·36°F	0·289°C / 0·52°F	0·04°C / 0·072°F

Tiers compared.

Examination of Posterior Region.

Right side. Left side.

4th tier higher in temperature than 3rd tier by 0·246°C.(0·442°F.)—0·302°C.(0·543°F.).
3rd tier higher in temperature than 4th tier in 3rd and 5th districts by 0·085°C.(0·153°F.).
4th tier higher in temperature than 5th tier by 0·284°C.(0·512°F.)—0·261°C.(0·47°F.).
5th tier higher in temperature than 6th tier by 0·135°C.(0·243°F.)—0·086°C.(0·155°F.).

The order of average supremacy of temperature of the respective tiers is very different on the two sides of the posterior region. The following is the order for each side:

Right side—
1st tier ; 2nd tier; 4th tier; 3rd tier ; 5th tier; 6th tier.
Left side—
4th tier; 1st tier; 2nd tier ; 5th tier; 6th tier ; 3rd tier.

Taking the average absolute temperature of the 1st tier, right side, as 33·676° C. (92·616° F.); and that of the same tier, left side, as 33·583° C. (92·449° F.), we have the following average absolute temperatures for each tiers of each side:

Right side. Left side
1st tier— 33·676° C. (92·616° F·)—33·583° C. (92·449° F·).
2nd „ —33·643° C. (92·557° F.)—33·540° C. (92·372° F.).
3rd „ —33·486° C. (92·274° F.)—33·336° C. (92·004° F.).
4th „ —33·600° C. (92·480° F.)—33.638° C. (92·548° F.).
5th „ —33·316° C. (91·968° F.)—33·377° C. (92·078° F.).
6th „ —33·181° C. (91·725° F.)—33·291° C. (91·923° F.).

If we compare the percentages of times of occurrence of superiority of temperature of one tier over another and the thermometric values of the differences of temperature between the tiers we have the following figures:

Temperature of the Head.

	Percentages of times of occurrence of superiority of temperature of one tier over another, on one and the same side of the head. Mean results of both sides taken together.	Mean differences of temperature between adjoining tiers, on one and the same side of the head. Mean results of both sides taken together.
1st tier superior to 2nd tier	64·000	0·107° C. (0·192° F.).
2nd tier superior to 1st tier	63·000	0·067° C. (0·12° F.).
2nd tier superior to 3rd tier	58·800	0·18° C. (0·328° F.).
4th tier superior to 3rd tier	63·600	0·274° C. (0·493° F.).
3rd tier superior to 4th tier	60·000	0·085° C. (0·153° F.).
4th tier superior to 5th tier	60·250	0·272° C. (0·489° F.).
5th tier superior to 6th tier	58·666	0·11° C. (0·198° F.).

We do not find, in a comparison of the above sets of values, any indication of a definite relation between the frequency of occurrence of superiority of temperature of one tier over another and the average degree of the difference of temperature observed.

The average difference of temperature for all the individual spaces, in the comparison by tiers, is, on the right side, 0·167° C. (0·3006° F.), and on the left side, 0·195° C. (0·351° F.). The average difference of temperature for the two sides taken together is, therefore, 0·181° C. (0·325° F.).

Comparison of spaces situated in two adjoining districts in the same tier, on one and the same side of the head.

TABLE 17.

Results of 50 observations on the comparative temperature of each pair of spaces situated in adjoining districts of the same tier on one and the same side of the posterior region. "Left" and "Right" signify, respectively, left and right sides of the head. The figures denote the numbers of times, in a total of 50, in which each space was superior in temperature to the space in the adjoining district of the same tier on the same side of the head:—thus, in fifty comparisons of the 1st and 2nd districts in the 1st tier, on the left side of the

Examination of Posterior Region.

head, the temperature was higher in the 1st district thirty times, and in the 2nd district twenty times.

Districts compared.

			1st & 2nd	2nd & 3rd	3rd & 4th	4th & 5th			
6th Tier	{	Right .	31	19	33	17			
		Left .	21	29	31	19			
5th Tier	{	Right .	16	34	29	21	21	29	
		Left .	20	30	34	16	30	20	
4th Tier	{	Right .	19	31	33	17	22	28	31	19
		Left .	17	33	28	22	21	29	30	20
3rd Tier	{	Right .	28	22	17	33	31	19	35	15
		Left .	32	18	29	21	32	18	21	29
2nd Tier	{	Right .	19	31	15	35	23	27	32	18
		Left .	29	21	16	34	30	20	34	16
1st Tier	{	Right .	22	28	19	31	30	20	32	18
		Left .	30	20	17	33	32	18	33	17

Analyzing the above table we obtain the following results:

1st. The 1st district is, in the majority of cases, of higher temperature than the 2nd district, in the 1st, 2nd, and 3rd tiers, on the left side, and in the 3rd and 6th tiers, on the right side. In the 1st and 2nd tiers, on the right side; in the 4th and 5th tiers, on both sides; and in the 6th tier on the left side, the 2nd district is, in the majority of cases, of higher temperature than the 1st district.

2nd. The 2nd district is, in the majority of cases, of higher temperature than the 3rd district, in the 3rd tier, on the left side, and in the 4th, 5th, and 6th tiers, on both sides. In the first and 2nd tiers on both sides, and in the 3rd tier, on the right side, the 3rd district is, in the majority of cases, of higher temperature than the 2nd district.

3rd. The 3rd district is, in the majority of cases, of higher temperature than the 4th district, in the 1st and 3rd tiers, on both sides, and in the 2nd and 5th tiers on the left side. In the 4th tier on both sides, and in the 2nd and 5th tiers, on the right side, the 4th district is, in the majority of cases, of higher temperature than the 3rd district.

4th. The 4th district is, in the majority of cases, of higher temperature than the 5th district in every tier, on both sides, with the exception of the 3rd tier tier left side, where the 5th district is, in the majority of cases, of higher temperature than the 4th district.

In estimating the mean percentages of times of occurrence of superiority of temperature of a given district over the district adjoining, it will be better, on account of the great irregularity which prevails in the position of the balance of superiority of temperature in different spaces of the same districts, to strike a balance between all the spaces of each pair of districts compared, and to let the result thus obtained stand for the whole of the two districts. Proceeding in this manner, we obtain the following values :

Percentages of times of occurrence of superiority of temperature.

	Right side.	Left side.
2nd district superior to 1st district . -	55,000 -	50,333
2nd district superior to 3rd district -	— -	51,666
3rd district superior to 2nd district -	51,333 -	—
3rd district superior to 4th district . -	50,800 ·	58,000
4th district superior to 5th district . -	65,000 -	59,000

Table 18 gives the thermometric values of the differences of temperature between adjoining districts. Analyzing the table we obtain the following results :

Average thermometric values of differences of temperature between adjoining districts of one and the same side of the posterior region.

	Right side.	Left side.
1st district higher in temperature than 2nd district by . .	—	0·003°C.(0.0054°F.).
2nd district higher in temperature than 1st district by . .	0·0272°C.(0·489°F.)	—
2nd district higher in temperature than 3rd district by . .	0·029°C.(0·052°F.)	—0·096°C.(0·172°F.).

Examination of Posterior Region.

TABLE 18.

Comparison in degrees Centigrade and Fahrenheit of each space of each district with the space in the district immediately adjoining of the same tier, on one and the same side of the posterior region. "Left" and "Right" signify, respectively, left and right sides of the head. The figures are placed in the district having the higher temperature of the two compared; thus, the figures 0·039° C. (0·0702° F.) in the left hand lowest space of the table, denote, that, on the left side of the head, in the 1st tier, the temperature is higher in the 1st district than in the 2nd district by these amounts. The results are the mean of 50 observations on each pair of spaces.

Districts compared.

Tier	Side	1st and 2nd.	2nd and 3rd.	3rd and 4th.	4th and 5th.
6th Tier	Right	0·065°C. 0·117°F.	0·208°C. 0·374°F.		
6th Tier	Left		0·017°C. 0·03°F.	0·273°C. 0·491°F.	
5th Tier	Right		0·013°C. 0·023°F.	0·255°C. 0·459°F.	0·034°C. 0·0612°F.
5th Tier	Left		0·035°C. 0·063°F.	0·22°C. 0·396°F.	0·04°C. 0·072°F.
4th Tier	Right		0·156°C. 0·28°F.	0·138°C. 0·248°F.	0·061°C. 0·109°F.	0·176°C. 0·316°F.
4th Tier	Left		0·143°C. 0·257°F.	0·113°C. 0·203°F.	0·088°C. 0·158°F.	0·17°C. 0·306°F.
3rd Tier	Right	0·065°C. 0·117°F.		0·386°C. 0·694°F.	0·032°C. 0·0576°F.	0·033°C. 0·0594°F.
3rd Tier	Left	0·016°C. 0·0288°F.	0·042°C. 0·075°F.		0·103°C. 0·185°F.	0·125°C. 0·225°F.
2nd Tier	Right	0·048°C. 0·086°F.		0·033°C. 0·0594°F.	0·08°C. 0·144°F.	0·144°C. 0·259°F.
2nd Tier	Left	0·161°C. 0·289°F.		0·048°C. 0·086°F.	0·064°C. 0·115°F.	0·064°C. 0·115°F.
1st Tier	Right		0·076°C. 0·136°F.	0·005°C. 0·009°F.	0·055°C. 0·099°F.	0·207°C. 0·372°F.
1st Tier	Left	0·039°C. 0·0702°F.		0·022°C. 0·0396°F.	0·241°C. 0·433°F.	0·106°C. 0·19°F.

Temperature of the Head.

	Right side.	Left side.
3rd district higher in temperature than 4th district by	—	0·072°C.(0·129°F.).
4th district higher in temperature than 3rd district by	0·0176°C.(0·031°F.)	—
4th district higher in temperature than 5th district by	0·14°C.(0·252°F.)	0·053°C.(0·0954°F.).

The order of superiority of absolute temperature of the districts is as follows:

Right side—
2nd district; 4th district; 1st district; 3rd district; 5th district.

Left side—
1st district; 2nd district; 3rd district; 4th district; 5th district.

Calling the average absolute temperature of the 1st district, right side, 33·494° C. (92·289° F.), and that of the same district, left side, 33·552° C. (92·393° F.), we have the following average absolute temperature for each district of each side :

		Right side.	Left side.
1st District—		33·494°C.(92·289°F.)	33·552°C.(92·393°F.)
2nd	,,	33·521°C.(92·337°F.)	33·549°C.(92·388°F.)
3rd	,,	33·492°C.(92·285°F.)	33·453°C.(92·215°F.)
4th	,,	33·509°C.(92·316°F.)	33·381°C.(92·085°F.)
5th	,,	33·369°C.(92·064°F.)	33·328°C.(91·9°F.)

We will next compare the percentages of times of occurrence of superiority of temperature of one district over another, with the degree of difference of temperature observed between the districts.

	Percentages of times of occurrence of superiority of temperature of one district over another on one and the same side of the head. Mean results of both sides taken together.	Mean difference of temperature between adjoining districts on one and the same side of the head. Mean results of both sides taken together.
1st district superior to 2nd district	47·333	0·0288° C.(0·0518° F.).
2nd district superior to 1st district	52·666	0·0406° C.(0·073° F.).

Examination of Posterior Region.

	Percentages of times, &c.	Mean differences, &c.
2nd district superior to 3rd district	50·166	0·104° C.(0·187° F.).
3rd district superior to 2nd district	49·834	0·041° C.(0·0738° F.).
3rd district superior to 4th district	54·400	0·053° C.(0·0954° F.).
4th district superior to 3rd district	45·600	0·0263° C.(0.0473° F.).
4th district superior to 5th district	62·000	0·096° C.(0·172° F.).

There is no evidence of any definite relation between the two classes of values compared above.

The average difference of temperature for all the individual spaces, in the comparison by districts, is, on the right side, 0·108° C. (0·194° F.); and on the left side, 0·101° C. (0·181°F.). The average difference of temperature for the two sides taken together is, therefore, 0·1045° C. (0·188° F.).

Lastly, taking the three principal classes of observation, as was done in the cases of the anterior and middle regions, we have the following values:

Comparison of the two sides of the head.

Average difference of temperature.

Average percentage of times of occurrence of superiority of temperature of either side of the head . . . . 66·00—0·115° C. (0·207° F.)

Comparison of adjoining tiers of one and the same side.

Average difference of temperature.

Average percentage of times of occurrence of superiority of temperature of one tier over another, on both sides taken together . . . . 60·545—0·181° C. (0·325° F.).

100   *Temperature of the Head.*

Comparison of adjoining districts of one and the same side.

Average percentage of times of occurrence of superiority of temperature of one district over another, on both sides taken together . . . . . . 62·047—0·1045° C. (0·188° F.).

Average difference of temperature.

From the above figures we learn that superiority of temperature of one side of the head over the other occurs more frequently than superiority of temperature of one tier or district, on one and the same side of the head, over another. Herein the posterior region differs from the anterior and middle regions. Like the two latter, the posterior region shows a greater percentage of times of occurrence of superiority of temperature in the comparison by districts than in that by tiers. With regard to thermometric values, the greatest difference of temperature is found in the comparison by tiers, and the least difference in the comparison by districts.

It now remains to group together for comparison the principal results obtained in the examination of all three regions. The following are the respective values selected for comparison:

Comparison of the two sides of the head.

Regions.	Percentages of times of occurrence of superiority of temperature.		Average differences of temperature.	
	Right side.	Left side.	Right side.	Left side.
Anterior	54·153	45·847	0·255°C. (0·459°F.)	0·241°C. (0·433°F.).
Middle	49·711	50·289	0·0589°C. (0·106°F.)	0·1103°C. (0·198°F.).
Posterior	45·458	54·542	0·186°C. (0·334°F.)	0·066°C. (0·118°F.).

From the above figures we learn that the anterior region shows the highest percentage of occurrence of superiority of temperature for the right side, and the lowest percentage for the left side: that the posterior region shows the highest percentage of occurrence of superiority of temperature for the left side, and the lowest per-

Comparison of the three Regions.

centage for the right side : that the middle region, on both sides, occupies an intermediate place with regard to frequency of occurrence of superiority of temperature, between the anterior and posterior regions.*

The anterior region shows the greatest mean difference of temperature on both sides : the posterior region, on the right side, shows the next greatest average difference; then come the middle region on the left side, and the posterior region on the same side: the smallest average difference is found in favour of the right side in the middle region.

Comparison of the two sides of the head.

Regions.	Greatest differences of temperature observed.	Smallest differences of temperature observed.	Range.
Anterior	0·461°C. (0·829°F.).	0·076°C. (0·136°F.).	0·385°C. (0·693°F.).
Middle	0·264°C. (0·475°F.).	0·016°C. (0·028°F.).	0·248°C. (0·446°F.).
Posterior	0·386°C. (0·694°F.).	0·008°C. (0·0144°F.).	0·378°C. (0·68°F.).

The greatest difference of temperature is found in the anterior region, and the second greatest difference in the posterior region. The smallest difference of temperature exists in the posterior region; the next in order, in this respect, is found in the middle region. The greatest range of difference of temperature belongs to the anterior region, and the least to the middle region.

Comparison of adjoining tiers of one and the same side.

Regions.	Percentages of times of occurrence of superiority of temperature of one tier over another ; both sides of the head taken together.	Average difference of temperature.
Anterior	66·045	0·21°C. (0·378°F.).
Middle	64·724	0·1555°C. (0·279°F.).
Posterior	60·545	0·181°C. (0·325°F.).

* Of the 8800 experiments tabulated, 4171 are in favour of the right side, 4222 in favour of the left side, and 407 in favour of equality of temperature.

102 *Temperature of the Head.*

In the comparison by tiers, the anterior region leads, both as regards frequency of occurrence of superiority of temperature and degree of difference of temperature. The middle region shows the next highest percentage of frequency of occurrence of superiority of temperature, and the least difference of temperature.

Comparison of adjoining districts of one and the same side.

Regions.	Percentages of times of occurrence of superiority of temperature of one district over another; both sides of the head taken together.	Average difference of temperature.
Anterior	66·119	0·055°C. (0·099°F.).
Middle	65·346	0·131°C. (0·235°F.).
Posterior	62·047	0·1045°C. (0·188°F.).

In the comparison by districts, the anterior region shows the highest percentage of times of occurrence of superiority of temperature, and the lowest difference of temperature. The middle region shows the next greatest percentage of occurrence of superiority of temperature, and the greatest difference of temperature.

In order to give an idea of the relative temperatures of the spaces of different regions, and likewise of spaces asymmetrically situated in one and the same region, Tables 19, 20, and 21, have been prepared. The values given in these tables are all referred by comparison to the value assigned in Table 20 to the 2nd district, 2nd tier, right side, of the middle region, namely, 34°C. (93·2° F.). From this last value as a standard, the thermometric degrees are estimated for the other spaces, partly by the differences of temperature given in preceding tables of the different regions, and partly by direct observations. It will frequently be found that the difference of temperature obtained by comparing two absolute values in Tables 19, 20, 21, is not the same as that given in the tables of the different regions. Whenever this occurs the values given in the latter class of tables should be accepted by preference, as representing the results of a larger number of observations. The tables of absolute values have been given principally to furnish *qualitative* not *quantitative* information.

Absolute Temperatures.

TABLE 19.—Absolute thermometric values of anterior region.

Tier	Right side 5th	Right 4th	Right 3rd	Right 2nd	Right 1st	Districts 1st	Districts 2nd	Left 1st	Left 2nd	Left 3rd	Left 4th	Left 5th
6th Tier										33.657°C / 92.582°F	33.75°C / 92.75°F	
5th Tier			33.9°C / 93.02°F						33.51°C / 92.318°F	33.63°C / 92.534°F	33.854°C / 92.937°F	33.645°C / 92.561°F
4th Tier			33.81°C / 92.858°F	33.65°C / 92.57°F	33.6°C / 92.48°F	33.5°C / 92.3°F	33.51°C / 92.318°F	33.475°C / 92.255°F	33.55°C / 92.39°F	33.701°C / 92.661°F	33.984°C / 93.171°F	34.015°C / 93.227°F
3rd Tier		33.84°C / 92.912°F	33.82°C / 92.876°F	33.5°C / 92.3°F	33.55°C / 92.39°F	33.748°C / 92.743°F	33.497°C / 92.294°F	33.5°C / 92.3°F	33.94°C / 93.092°F	33.952°C / 93.113°F	34.034°C / 93.261°F	34.035°C / 93.263°F
2nd Tier	33.781°C / 92.805°F	33.955°C / 93.119°F	33.841°C / 92.913°F	33.775°C / 92.795°F	33.748°C / 92.743°F	33.759°C / 92.766°F	33.86°C / 92.948°F	33.86°C / 92.948°F	34.00°C / 93.2°F	34.032°C / 93.257°F	33.904°C / 93.027°F	32.825°C / 92.885°F
1st Tier	34.141°C / 93.453°F	34.151°C / 93.471°F	33.931°C / 93.075°F	33.8°C / 92.84°F	33.759°C / 92.766°F	33.775°C / 92.795°F	33.92°C / 93.056°F	33.81°C / 92.858°F		33.884°C / 92.991°F		
(1st Tier cont.)	34.161°C / 93.489°F	34.181°C / 93.525°F	33.781°C / 92.805°F	33.921°C / 93.056°F	33.775°C / 92.795°F							
(1st Tier cont.)	34.131°C / 93.435°F	34.081°C / 93.345°F		33.756°C / 92.76°F								

Temperature of the Head.

TABLE 20.—Absolute thermometric values of middle region.

	Right side.					Left side.				
	Districts.									
	5th.	4th.	3rd.	2nd.	1st.	1st.	2nd.	3rd.	4th.	5th.
7th Tier	32·845°C. 91·112°F.	32·862°C. 91·151°F.	32·887°C. 91·196°F.	33·126°C. 92·919°F.	33·307°C. 91·952°F.	33·258°C. 91·864°F.	33·192°C. 91·745°F.	33·101°C. 91·581°F.	33·018°C. 91·432°F.	32·96°C. 91·328°F.
6th Tier	32·96°C. 91·328°F.	32·977°C. 91·358°F.	32·994°C. 91·389°F.	33·193°C. 91·747°F.	33·422°C. 92·159°F.	33·34°C. 92·012°F.	33·274°C. 91·893°F.	33·175°C. 91·715°F.	33·158°C. 91·684°F.	33·117°C. 91·61°F.
5th Tier	32·763°C. 90·973°F.	32·944°C. 91·299°F.	33·282°C. 91·907°F.	33·48°C. 92·264°F.	33·529°C. 92·352°F.	33·414°C. 92·145°F.	33·562°C. 92·411°F.	33·546°C. 92·382°F.	33·109°C. 91·596°F.	32·779°C. 91·022°F.
4th Tier	33·224°C. 91·803°F.	33·74°C. 91·893°F.	33·356°C. 92·04°F.	33·406°C. 92·13°F.	33·546°C. 92·382°F.	33·48°C. 92·264°F.	33·447°C. 92·204°F.	33·389°C. 92·1°F.	33·34°C.	33·249°C. 91·848°F.
3rd Tier	33·29°C. 91·922°F.	33·406°C. 92·13°F.	33·505°C. 92·309°F.	33·711°C. 92·679°F.	33·777°C. 92·798°F.	33·727°C. 92·708°F.	33·645°C. 92·561°F.	33·422°C. 92·175°F.	33·373°C. 92·071°F.	33·274°C. 91·893°F.
2nd Tier	33·323°C. 91·981°F.	33·422°C. 92·159°F.	33·554°C. 92·397°F.	34·00°C. 93·2°F.	34·115°C. 93·407°F.	34·082°C. 93·347°F.	33·918°C. 93·052°F.	33·521°C. 92·337°F.	33·406°C. 92·13°F.	33·307°C. 91·952°F.
1st Tier	33·241°C. 91·833°F.	Ear	33·844°C. 92·919°F.	34·024°C. 93·243°F.	34·132°C. 93·437°F.	34·165°C. 93·497°F.	33·934°C. 93·081°F.	33·786°C. 92·814°F.	Ear	33·043°C. 91·477°F.

Absolute Temperature.

TABLE 21.—Absolute thermometric values of posterior region.

	Right.					Districts.					Left.	
	5th.	4th.	3rd.	2nd.	1st.	1st.	1st.	2nd.	3rd.	4th.	5th.
6th Tier			33·085°C. 92·93°F.	33·293°C. 91·927°F.	33·358°C. 92·044°F.	33·438°C. 92·188°F.	33·455°C. 92·219°F.	33·182°C. 91·727°F.		
5th Tier		33·249°C. 91·848°F.	33·215°C. 91·787°F.	33·47°C. 92·246°F.	33·457°C. 92·222°F.	33·495°C. 92·291°F.	33·53°C. 92·354°F.	33·31°C. 91·958°F.	33·27C° 91·886°F.	
4th Tier	33·472°C. 92·249°F.	33·648°C. 92·566°F.	33·587°C. 92·456°F.	33·725°C. 92·705°F.	33·569°C. 92·424°F.	33·59°C. 92·462°F.	33·733°C. 92·719°F.	33·62°C. 92·516°F.	33·708°C. 92·674°F.	33·538°C. 92·368°F.
3rd Tier	33·582°C. 92·447°F.	33·615°C. 92·507°F.	33·647°C. 92·564°F.	33·261°C. 91·789°F.	33·326°C. 91·986°F.	33·39°C. 92·102°F.	33·374°C. 92·073°F.	33·332°C. 91·997°F.	33·229°C. 91·812°F.	33·354°C. 92·037°F.
2nd Tier	33·599°C. 92·478°F.	33·743°C. 92·737°F.	33·663°C. 92·593°F.	33·63°C. 92·534°F.	33·582°C. 92·447°F.	33·679°C. 92·622°F.	33·518°C. 92·332°F.	33·566°C. 92·418°F.	33·502°C. 92·303°F.	33·438°C. 92·188°F.
1st Tier	33·495°C. 62·291°F.	33·702°C. 92·663°F.	33·757°C. 92·762°F.	33·752°C. 92·753°F.	33·676°C. 92·616°F.	33·719°C. 92·594°F.	33·68°C. 92·624°F.	33·702°C. 92·693°F.	33·461°C. 92·229°F.	33·355°C. 92·039°F.

Temperature of the Head.

If the three regions be compared with each other, the following figures, obtained from the above tables, will represent approximately, their relative thermal values expressed in absolute quantities:

Averages of temperature.

	Anterior region.	Middle region.	Posterior region.
Right side	33·857°C.	33·374°C.	33·524°C.
	(92·942°F.).	(92·073°F.).	(92·343°F.).
Left side	33·792°C.	33·397°C.	33·487°C.
	(92·825°F.).	(92·114°F.).	(92·267°F.).
Both sides	33·824°C.	33·785°C.	33·505°C.
	(92·883°F.).	(92·093°F.).	(92·309°F.).

Extremes of temperature.

	Anterior region.	Middle region.	Posterior region.
Highest temperature	34·181°C.	34·165°C.	33·757°C.
	(93·525°F.).	(93·497°F.).	(92·762°F.).

Locality.—4th district 2nd tier, right side; 1st district 1st tier, left side; 3rd district 1st tier, right side.

Lowest temperature	33·475°C.	32·763°C.	33·085°C.
	(92·255°F.).	(90·973°F.).	(91·553°F.).

Locality.—2nd district 5th tier, left side; 5th district 5th tier, right side; 3rd district 6th tier, right side.

It will be seen from the last two sets of values that the anterior region has the highest, and the middle region the lowest average temperature; that the highest individual temperature occurs in the anterior region, and the second highest in the middle region; that the lowest individual temperature occurs in the middle region, and the second lowest in the posterior region.

It must not, however, be supposed that the above figures actually represent the absolute temperatures usually met with on the surface of the head. The temperatures given in the tables—particularly in the case of the anterior region—are such as would be found under the most favorable experimental conditions, where all precautions had been exercised to ensure perfect mental and bodily quietude, and where the physical surroundings were fully under control. Higher values, especially for the anterior region, are more frequently met with in ordinary examinations; but it is extremely difficult to arrive at satisfactory conclusions as to average absolute temperatures from observations made upon individuals while engaged in the ordinary avocations of life, the results thus obtained being very variable. Under such circumstances, a temperature of 35·2° C. (95·36° F.) for the anterior region; one of 34·5° C. (94·1°F.); for the middle region; and one of 34·2° C. (93·56° F.), for the posterior region, would probably represent more correctly, in the majority of cases, the highest absolute temperatures than the figures given above in the final set of values.

CHAPTER VII.

CAUSES OF DISTURBANCE AFFECTING THE EXPERIMENTS.—CONSIDERATIONS REGARDING THE INFLUENCE OF THE TEMPERATURE OF THE BRAIN ON THE TEMPERATURE OF THE OUTER SURFACE OF THE HEAD.—PROPAGATION OF SLIGHT DIFFERENCES OF TEMPERATURE IN BAD CONDUCTORS.—EFFECT OF THE CIRCULATION OF THE BLOOD ON THE OUTWARD TRANSMISSION OF HEAT FROM THE BRAIN TO THE INTEGUMENT.—GENERAL CONCLUSIONS REGARDING A CORRESPONDENCE BETWEEN THE RELATIVE TEMPERATURES OF POINTS OF THE INTEGUMENT AND THE RELATIVE TEMPERATURES OF UNDERLYING POINTS OF THE BRAIN SURFACE.

HAVING thus examined in turn each of the three regions of the head, it will be well to attend, for a moment, to a subject which concerns all the investigations which we have been considering. It is the avoidance of certain causes of disturbance which are liable to interfere with the accuracy of our results. Allusion has already been made to these sources of error on p. 28. They may be divided into two classes, external and internal. The most common and troublesome of the external causes are:

1st. Changes in the temperature of the air.

2nd. Exposure of the head to the rays of the sun, to radiation from fires or heated surfaces, and to the wind or to draughts of air.

Changes in the temperature of the air in which the examination of the head is made can usually be limited so as to produce no effect upon the temperature of the body; but the effect frequently produced by sudden transition from a cooler to a warmer, or from a warmer to a cooler atmosphere, is of a more troublesome character. The author has so far been unsuccessful in his attempts to establish any order or rule for the disturbances of temperature in the head which occur under these circumstances. Moreover, these disturbances often persist for some time. If, then, there be much difference between the temperature of the atmosphere to which the subject of the examination has been previously exposed and the temperature of the room in which the experiment is made, either a sufficient time must be allowed to elapse before commencing the observation, or the results obtained must be accepted with reservation.

With regard to the second class of causes of disturbance, very confusing and contradictory variations of natural temperature are observed. The head may be so placed as to be acted upon unequally in different parts, or, the exposure of two parts being precisely the same, one part may be, for some unknown reason, more affected than the other. The results obtained under these circumstances, like those mentioned above, appear to follow no definite rule or order. The effects are sometimes slight and transient in character, and at other times marked and persistent, and this, too, when the intensity of the cause appears to be the same.

The internal disturbing causes may be divided into three classes, namely:

1st. Alterations of the general circulation.
2nd. Alterations of local circulation.
3rd. Alterations of the degree of mental activity.

The usual effect of an increase in the activity of the general circulation is, as would be expected, to increase the percentage of neutrality by favouring the establishment of thermal equality between points of different temperatures. Accordingly, we usually see, with a general increase of absolute temperature, a diminution in the degree of difference of temperature of different points; but this is by no means invariably the case. In one individual examined by the author, increased activity of the general circulation usually brought about a greater difference in the distribution of temperature in the middle region than existed before, the extent of the tract of superior temperature for the left side being increased. In another case, the balance of higher temperature in several of the spaces of the inner districts of the anterior region generally shifted over from the left to the right side, thus increasing the extent of tract already in favour of the latter. No order can be assigned to these disturbances, which are different in different individuals, and which soon disappear as the circulation quiets down; but they inculcate a useful lesson as to the circumstances under which the head should be examined for the first time. If the head has been previously studied, these disturbances are easily recognised and accounted for; but, in examining a head for the first time, the experiment should be made, if possible, in the morning, before the individual has entered upon any active employment. A second examination, made a few hours later in the day, will often develop results very different from those obtained at the earlier period.

Alterations of local circulation are of common occurrence, especially in the anterior region, these alterations, under the form of irregular changes in the calibre of the blood-vessels, constituting flushing of the surface, are often exceedingly troublesome. The effects produced appear to follow no rule applicable to all cases, but in two individuals carefully examined in this respect by the author, the increase of temperature observed was always more marked on a single side—in one case on the right side, and in the other case on the left side. Sometimes the tract of surface affected is small, while at other times the greater part of a region may be involved. All that can be done when these conditions exist is to abandon the observation, as the results obtained under such circumstances are wholly unreliable.

The state of the mind must be carefully attended to in examining the relative temperatures of the different parts of the surface of the head. We shall study in full, in Part III, the effect of mental changes on the temperature of the head, it is, therefore, only necessary to say here that, mental excitement, both intellectual and emotional, must be carefully avoided, both at the time of the examination and also for a little time previous. In this connection it should be borne in mind that the mere arousing of the attention suffices, in some individuals, to bring about an elevation of temperature in the head, as has been mentioned in a former place (Introduction, p. 3). The watching of the spot of light of a reflecting galvanometer will frequently, in persons seeing the instrument for the first time, produce a rise of temperature.

Besides the disturbing causes given above, the position of the head may modify the temperature of its surface. If the subject of the examination has been lying upon one side of the head, or resting one side of the head upon the hand, the side thus lain or rested upon will usually be found to have its temperature augmented.

Alcoholic liquors in some persons, even when taken in great moderation, appear also to alter the normal temperature, and that, too, often in a decided degree. In some of these cases the effect is simply to confuse the usual order of distribution of temperature, the rearrangement following no fixed rule. In one case a very small quantity of alcohol (half a glass of weak ale) almost always caused the balance of superiority of temperature, in the anterior region, to change from the right side to the left, in some instances leaving only three or four spaces in favour of the right side; there was, however, no regularity as to the particular spaces the temperatures of which were affected.

We have now brought to a conclusion the first of the two parts into which our investigations are divided. We have examined in detail the normal relative temperatures of the different portions of the surface of the head, taken in definite subdivisions, and have obtained thermal averages for these subdivisions. The question now arises, To what extent are these different temperatures observed at the surface dependent upon differences of temperature of points of tissues underlying the skin, notably, the surface of the brain? This question is by no means an easy one to answer; in fact, a satisfactory reply is, in our present state of knowledge, impossible. To begin with, it is not for a moment to be supposed that the temperature of any given point at the surface of the head is under the immediate control of the temperature of the portion of cerebral tissue lying directly beneath it, and that, therefore, a difference in temperature of two points on the outer surface necessarily implies a difference of the same kind between the underlying corresponding points of the surface of the brain. Such a theory is manifestly untenable.

Leaving out for a time the influence of the circulation of the blood, the temperature of any isolated point at the surface of the head must be the resultant of its own temperature and of the different temperatures of all the parts lying beneath it. For the sake of illustration, imagine a cylinder cut out in the head, having for one end the surface of the brain, and for the other end the skin. Further, let us consider the different layers of tissue as so many separate heat-producing centres. We shall then have—proceeding from the surface inward—skin, muscle, bone, membranes, and brain as so many furnaces of different powers. A change in the rate of calorific production of one of these furnaces will affect the temperature of the others simply by conduction, and how far each one will be affected will depend merely upon its distance from the centre of increased or diminished activity, and upon the conductivity of the intervening substance.* One part might increase, permanently and considerably, its production of heat, and the other parts might have their temperatures raised, and yet, by reason of low conductivity, never arrive at thermal equality with the first part, when better conductors would all attain the same temperature.

Now, the brain has the highest temperature of all the tissues concerned in our cylinder, consequently, its temperature must, in the end, be the main factor in the sum total of all the different temperatures, and an increase or a diminution in its rate of production of heat would have a greater effect upon the temperatures of all the parts concerned than a proportionate change in the rate of production of any other one part.

In the next place, suppose that we have two similar cylinders, the two portions of brain-tissue of which produce unequal amounts of heat, the only difference existing between the two cylinders is the greater production of heat in one of the furnaces of one cylinder than in the corresponding furnace of the other. To what extent this extra production of heat may influence perceptibly the temperature of a distant point—the outer surface of the skin for instance—will depend upon the degree of increased production, and upon the conductivity of the intervening tissues. The solution of this part of the problem is, therefore, to be found in an examination of the rate of propagation of heat through bad conductors, like the animal tissues.

That the animal tissues are very poor conductors there is no doubt. The mere fact that too closely contiguous points of the surface of the body will maintain different temperatures, even

* We are here supposing the different tissues to have equal specific heats.

when well protected from all external thermal influences, is sufficient proof of the low power of conduction of these tissues; but it is no contradiction to say, that, although the animal tissues are such bad conductors that points very near each other may not come into a state of thermal equality, yet the heating or cooling of one point may influence the temperature of the other, the question being merely one of the amount of heat transmitted from one point to the other per unit of time; if the rate of propagation be not sufficiently high, thermal equilibrium will not be established, although the temperature of one point may still be affected by that of the other.

To ascertain the rate of propagation of small differences of temperature through badly conducting substances, the author made a series of experiments upon paraffin, spermaceti, beeswax, lard, and butter. The general method of experimenting was as follows:

When the substance employed was solid, it was cut into a cake of the desired thickness, and of a sectional area considerably larger than the face of the pile to be employed in the testing. To one surface of the cake the pile was closely and firmly attached by warming this surface. The whole of the pile, not in contact with the cake, was then thickly enveloped in cotton wool soaked in paraffin, the coverings extending laterally beyond the borders of the cake, so as to prevent external changes of temperature from finding any shorter road to the pile than through the thickness of the cake to the surface in contact with the face of the pile. When semisolid substances were used, they were placed in a wooden cylinder closed at one end with a thin copper plate, and the face of the pile, introduced into the open end of the cylinder, was brought in contact with the substance. The pile was then protected as in the previous case.

The pile being connected with the galvanometer, when the latter indicated no current, the face of the pile, and therefore the surface of the cake in contact with it, would be at the temperature of the air. This latter value being known, the free surface of the cake, or the copper end of the wooden cylinder, was brought in contact with water or oil having a temperatnre differing in the desired degree from that of the air. The movement of the needle which ensued marked the gradual steps towards equalisation of temperature of the two surfaces of the cake, or of the substance contained in the cylinder. The follow-

Thermal Transmission from Brain to Integument. 113

ing is a fair example of a large number of experiments made on the rate of propagation of heat through paraffin. The cake of paraffin was 20 mm. (0·78 inch) in thickness, and the difference of temperature between its two surfaces was, at the start, 0·125° C. (0·225° F.). The figures of the second column are galvanometric degrees, the value of one degree being, in this instance, equal to 4·45° of the Thomson scale at 1000 mm. distance. The figures of the third column are the thermometric values of the deflections of the galvanometer. The temperature of the air was 10·125° C. (50·225° F.), and that of the water was 10° C. (50° F.). The transmission of heat in this case was therefore from the pile to the liquid.

		Deflections of needle.	Thermometric values.
At the commencement of the experiment	. .	0°	0°.
At the end of 0·75 of a minute		0·25°	0·0025°C. (0·0045°F.).
,,	1·25 minutes	0·5°	0·005°C. (0·009°F.).
,,	2·00 ,,	1·5°	0·015°C. (0·027°F.).
,,	3·00 ,,	2·5°	0·025°C. (0·045°F.).
,,	4·00 ,,	2·75°	0·0275°C. (0·0495°F.).
,,	5·00 ,,	3·50°	0·035°C. (0·063°F.).
,,	6·00 ,,	4·25°	0·0425°C. (0·0765°F.).
,,	7·00 ,,	5·25°	0·0525°C. (0·0945°F.).
,,	8·00 ,,	6·50°	0·0650°C. (0·117°F.).
,,	9·00 ,,	7·25°	0·0725°C. (0·1305°F.).
,,	10·00 ,,	7·50°	0·0750°C. (0·135°F.).
,,	11·00 ,,	8·00°	0·08°C. (0·144°F.).
,,	12·00 ,,	8·50°	0·085°C. (0·153°F.).
,,	13·00 ,,	8·75°	0·0875°C. (0·1575°F.).
,,	14·00 ,,	9·00°	0·09°C. (0·162°F.).
,,	15·00 ,,	9·25°	0·0925°C. (0·1665°F.).
,,	16·00 ,,	9·50°	0·095°C. (0·171°F.).
,,	17·00 ,,	,,	,, ,,
,,	18·00 ,,	,,	,, ,,
,,	19·00 ,,	,,	,, ,,
,,	20·00 ,,	,,	,, ,,

Thus in three quarters of a minute a difference of temperature of 0·125° C. (0·225° F.) made itself manifest by conduction through 20 mm. (0·78 inch) of paraffin; and at the end of sixteen minutes, at which time the permanent thermal condition was reached, the difference of temperature between the two surfaces of the paraffin cake was reduced to $0·03°$ C. ($0·054°$ F.); that is to say, the rate of transmission of heat through the cake was now more than three quarters of the difference of rate of production of heat at the two surfaces.

This experiment indicates not only, that, under circumstances such as we have supposed on p. 111, so far as the question of mere conductivity is concerned, the temperature of the brain can make itself felt as an important factor in the temperature of the surface of the head, but also, that, with sufficiently delicate apparatus, a slight change of temperature at the surface of the brain may be quickly detected at the exterior of the head.

Our two cylinders, the inner furnaces of which produce different amounts of heat, might, therefore, even when the difference of production was slight, exhibit unequal temperatures in their outer layers. But we have supposed the two cylinders to be precisely similar in size and in composition, and to be each isolated. If, however, the cylinders be in contact with each other, or be of different lengths, or, again, be composed of different substances, we have a much less simple condition of things to deal with than that which we have been considering.

In the first place, when the cylinders are in contact with each other, a portion of the extra heat produced in one, instead of being transmitted directly onward to the surface of that cylinder, is diverted to the second cylinder, raising the temperature of this latter, so that the outer layer of this cylinder has its temperature raised as well as the outer layer of the first cylinder, though in a less degree; hence, under these circumstances, a diminution of the difference of temperature noticed at the outer surface between the two cylinders, when isolated, will occur.

In the second place, if the cylinders be of different lengths, although the brain-tissue portion of one may produce more heat than that of the other, yet, if this latter cylinder be the shorter, its outer layer may receive more heat from the interior per unit of time than the first cylinder, and thus, of the two points, the temperatures of which are compared at the exterior, the one overlying the portion of brain producing the lesser amount of heat may exhibit the higher temperature.

In the third place, if the composition of the cylinders be not the same, the result will be a difference in their conducting powers; and greater conductivity may more than compensate for smaller production of heat; thus again, as in the last instance, causing the point of surface corresponding to the cooler portion of brain to manifest the higher temperature.

Now in examining the relative temperatures of different points of the surface of the head, we may have to deal with any one or all of the above conditions. The points compared may adjoin each other; the tissues intervening between the exterior and the surface of the brain may be of unequal thickness in different places; and the proportions of the different kinds of tissues may not be the same in all the points compared.

It is clear that symmetrically situated spaces of the two sides of the head are less liable to be affected by the last two sources of error than asymmetrically situated spaces, and it would therefore seem that—other things being equal—so far as the question of mere conductivity is concerned, the averages obtained in the experiments in which spaces on one and the same side were compared are less likely to represent correctly the thermal relations of the portions of cerebral tissue underlying the points examined on the outer surface, than the averages obtained in comparing symmetrically situated spaces of the two sides.

But the other tissues composing our cylinders may also alter their respective rates of production of heat as well as the brain tissue. Leaving out the special effects of nervous influence— vaso-motor or thermogenic—and regarding only differences in the activity of the production of heat in different points, existing independently of outside influences, it is evident that if one of the tissues, other than the brain, of one of the cylinders produce heat in a greater ratio than the corresponding tissue of the other cylinder, the true thermal relations of the brain-tissue portions of the cylinders may be masked, and the temperatures observed at the surface not represent the relative temperatures of the underlying portions of cerebral substance.

We have thus far left out of consideration the influence of the blood circulating in the tissues overlying the brain upon the transmission of heat outward from the cerebral surface to the exterior of the skin. Now a portion of the heat produced in the brain, over and above that produced in the tissues lying outside of it, instead of being directly transmitted to these tissues, must be expended in raising the temperature of the blood circulating between the brain and the exterior of the head.* It has been thought that this consumption of heat by the blood would effectually check thermal propagation outward, and thus prevent the temperature of the surface of the brain from directly influencing the temperature of the exterior of the head; or, admitting that all the excess of production of heat on the part of the brain over and above the production on the part of the exterior tissues was not expended upon the blood, but was in part transmitted to the outer surface, still, that this transmission was so insignificant that small differences in the rate of thermal production in the brain could not manifest themselves at the exterior surface.

Suppose now a cylinder composed as before described, but having each layer of tissue permeated by a set of tubes, which latter are constantly receiving and transmitting onward a liquid slightly cooler than the tissues which it traverses. Commencing with the first layer interiorly—the surface of the brain—a given volume of liquid arriving receives from the cerebral substance a certain amount of heat, which, in the absence of the liquid, would have been directly transmitted by conduction to the second layer of tissue. If the brain tissue and liquid possess the same specific heats, the result of the contact of the liquid with the warmer tissue will be to immediately raise the temperature of the former; but the moment this elevation of temperature takes place, the liquid will, in turn, at once lose a portion of this newly acquired heat to the cooler liquid and tissue of the second layer of the cylinder. No matter how rapidly the liquid may move this acquisition and subsequent loss must inevitably occur. Rapidity of movement of the liquid, by distributing the heat received in each unit of time through a large number of molecules, diminishes the effect produced upon each individual molecule, and this, in turn, diminishes the effect which each molecule can produce upon others, in its neighbourhood, cooler than itself; but there is no such thing as a removal of heat—as if it were a material substance —from the place of its production, with such rapidity as to pre-

* Were *all* the excess of production of heat in the brain, over and above the production in the exterior tissues, consumed in the blood, the higher temperature of the brain would be due simply to the greater distance of the organ from the outer surface, and consequent diminution of loss of heat exteriorly.

vent its affecting neighbouring parts. The transmission of heat outward in the cylinder is, therefore, diminished but not abolished, by the passage through the latter of the liquid.

What is true of the first layer of the cylinder will hold good of each successive layer, only the sum total of loss of heat will increase. If our cylinder be not isolated, although surrounding parts may receive from the liquid a certain amount of heat at the expense of the immediate locality of production, yet this point will still be the main focus of heat, the liquid will still be hottest here, and neighbouring liquids and tissues will still—other things being equal—receive a larger share of the surplus heat than more distant parts.

But considering the blood and brain tissue as of different specific heats, the blood being in this respect superior to the tissue, the loss of heat in warming the circulating fluid would be greater than under the circumstances just considered. In these new conditions a portion of the heat communicated by the brain tissue would be expended in interior work in the blood, and would be lost as thermometric heat. Now supposing the specific heat of the blood to be twice that of the brain tissue, the latter would lose, in warming the blood to the same degree as in the preceding instance, twice as much heat as was then lost. We are, however, in no position to estimate accurately the amount of heat which would be expended in warming the blood or in satisfying its capacity for heat. What we know is, that the cerebral substance produces sufficient heat to both satisfy the capacity for heat of the incoming blood, and to raise the temperature of the latter, since the temperature of the blood passing from the sinuses of the dura mater into the internal jugular vein is higher than that of the blood entering by the carotid artery. The brain is, in fact, hotter than the blood supplying it. The blood does not, therefore, reduce the brain to its own temperature, but, on the contrary, is itself heated; and this being the case, there must be a transmission outward of a portion of this surplus heat. The temperature of the cerebral substance must, therefore, in spite of the heat lost to the blood, exercise an important influence on the temperature of the exterior of the head.*

* With regard to the propagation of heat outward, it should be remarked that the capillary networks may, by their arrangement, cause the same blood to pass *more than once* over a given focus of heat, thus diminishing the loss of heat at this point.

But although the circulation cannot prevent the temperature of the brain from forming an important factor in the temperature of the surface of the head, it may still be asked if this circulation will not, by its tendency to produce thermal equilibrium between different points, effectually prevent the establishment of differences of temperature in neighbouring localities sufficient to manifest themselves at the outer surface. To this question it may be replied that if the circulation could really remove the excess of heat produced at a given point, over and above that produced at adjacent points, as rapidly as it is formed, it would be difficult to understand how in very vascular tissues, like the internal layer of the prepuce, the glans penis, and the lips, points situated in close proximity to each other, can maintain different temperatures, because, in these localities, two foci of heat of different powers ought to come to a state of thermal equilibrium through the medium of the close vascular network connecting them. Such, however, is not the case, for it is easy to show that, in the parts in question (due precautions having been taken to protect them from external cooling or heating influences), differences of temperature exist in points situated close to each other, as well as in less vascular tissues, the difference between the two classes of cases being simply one of *degree*. Moreover, it requires but little experience in the use of thermo-electric needles in the examination of the relative temperatures of points beneath the surface, to know that differences exist here as well as on the skin, even when the distance between the needles is slight. Again, the fact that the temperature of points of the surface over large arteries is usually higher than that of neighbouring points is contradictory of the idea that the circulation over a focus of heat can mask the position of such focus by distributing the heat to other parts; for, in these cases, the vascular networks of the tissues lying over the arteries ought to bring about equalisation of temperature in points directly above, and in those at a little distance from, the source of thermal supply.

The final question assumes really this form:—Can a slight increase of temperature at the surface of the brain, maintained for a certain length of time, affect perceptibly the outer surface of the head, after making reasonable allowances for imperfect conductivity and for loss of heat to the blood?

If it be granted that such a change of temperature can make

itself felt externally, it then follows that slight differences of temperature between points of the brain surface may also affect the temperature of the outer surface of the head.

To answer the above question, we will suppose the increase of temperature at the surface of the brain to be $0.1°$ C. ($0.18°$ F.); that one half of this is lost in warming the blood; and that the remainder encounters the same resistance that would be offered by 20 mm. of paraffin. We have, then, $0.1°$ C. reduced, first, to $0.05°$ C. ($0.09°$ F.), and afterwards, still farther reduced (by 24 per cent.), in its passage through the tissues, to $0.038°$ C. ($0.0684°$ F.), which latter value would, therefore, represent, at the outer surface of the head, the increase of $0.1°$ C. at the surface of the brain. Such a calculation is certainly not made too favorable to the possibility of the transmission of changes of temperature from the brain to the exterior surface.

It will be seen, however, that although reason has been given to believe, first, that the brain—in spite of the non-conductivity of the tissues and the influence of the circulation—is the principal factor in the temperature of the exterior of the head; and, second, that small differences of temperature at the surface of the brain may be detected at the outer surface of the head, yet there is no certainty that the different relative temperatures observed at the exterior surface represent correctly, either in kind or in degree, the relative temperatures of the corresponding underlying points of cerebral tissue. The question which was propounded at the outset (p. 110) is consequently by no means yet fully answered. We must, in our present state of knowledge, be governed simply by probabilities. It would, however, seem somewhat more than probable that the mean relative temperatures of tracts of the external surface, larger than those included in the subdivisions adopted in this work, bear definite relations to the mean relative temperatures of the tracts of brain surface lying beneath: but further and special investigations on this point are required. Meanwhile,—bearing in mind all the uncertainties and difficulties which have been mentioned,—the author does not feel inclined to place much reliance on the examination of the relative temperatures of different parts of the integument of the head as a means of medical diagnosis, although such examinations may in some cases lead to interesting results.

PART III.

The effect of Intellectual and Emotional Activity on the Temperature of the Head.

CHAPTER I.

GENERAL CONSIDERATIONS.—EFFECT OF DIFFERENT KINDS OF INTELLECTUAL WORK ON THE TEMPERATURES OF THE THREE REGIONS.

We have now to deal with the second of the two main objects of our investigations, namely, the changes of temperature produced at the exterior surface of the head by changes in the degree of production of heat in the cerebral substance, accompanying alterations in the degree of psychical activity.

We have seen* that there is good reason to suppose that a slight change of temperature at the surface of the brain can show itself, in a degree readily perceptible by means of delicate apparatus, at the exterior surface of the head; and we shall see, in the experiments now to be given, that the changes of temperature observed in the integument during increased mental activity are usually of a degree which may be fairly accounted for by the direct propagation outward, through the intervening tissues, of slight thermal changes at the cerebral surface.

The experiments which we have to consider may be included under three general heads, namely :

1st. Those which deal with the absolute effect of mental activity on the temperature of different parts of the surface of the head, without special reference to exact position.

* P. 119.

General Considerations.

2nd. Those which deal with the comparative effect of mental activity on the temperature of different points of one and the same side of the head.

3rd. Those which deal with the comparative effect of mental activity on the temperature of symmetrically situated points of the two sides of the head.

The brain work to be studied may be divided into two classes, namely :

1st. Intellectual activity.   2nd. Emotional activity.

Before proceeding to explain the experimental methods adopted, it will be well to consider certain psychological points connected with the rise of temperature during mental work.

To begin with intellectual work, although in every brain the performance of such work may be supposed to be attended with thermal changes, yet the readiness with which these changes may be detected varies very much in different individuals, and in the same individual with different kinds of work, even when, abstractedly considered, these different kinds may present equal difficulties of execution. In the first place, two conditions are usually essential to the production of heat in a degree sufficient to show itself distinctly at the outer surface of the head, namely : first, that the work be continuous for some minutes ; and, second, that, during this time, the mind be applied with a considerable degree of intensity. Either condition *may*, of itself alone, alter perceptibly the temperature, but generally, to produce a marked effect, both conditions are required. Thus prolonged application, even when such application has been at no one time intense, may bring about an elevation of temperature at the surface. And again, intense application for a very short time may produce the same result. The two conditions combined are, however, much more effective.

It is generally difficult and often impossible to detect any change of temperature in the heads of persons who work *leisurely*. The amount of work performed may be great, and, therefore, the total amount of heat produced may be considerable ; but the amount of work accomplished, and consequently the amount of heat produced, in each unit of time, is so small that the heat is lost in raising the temperatures of the blood and tissues nearest to the centre of production, and never makes itself felt at the outer surface.

Assuming that, in a given brain, the performance of a given amount of work of a certain kind and the amount of heat evolved during the work have a certain definite relation, it is easy to see that the *time* occupied in the accomplishment of the work becomes a highly important element of the question of degree of rise of temperature. If the work be concentrated into a short space of time, the evolution of heat per unit of time will be proportionally great. Thus, to take the simplest case, the addition of certain figures will represent a certain expenditure of force in a given brain, whether such addition be slowly or rapidly performed, and the total production of heat will be the same in both cases; but, in the slow addition, the production of heat per unit of time will be too small to be detected, it being lost, in the way above mentioned, as rapidly as it is produced.

The kind of work performed to bring about a rise of temperature in the head must vary with the individual. It should be of a nature either to present some considerable difficulty in its accomplishment, or to decidedly interest the individual. For example, persons who are not in the habit of expressing their ideas in writing may be made to raise the temperature of the head by earnestly endeavouring to write upon a given subject, even when the latter has been in the mind for some time and has been fully discussed, the work being thus limited to the simple composition. Not so, however, with the brain accustomed to composition, even when the subject written upon is a difficult one. In like manner, no effect is produced in persons accustomed to mathematical work by giving them problems to solve or sums to compute, unless either the task be one of unusual difficulty, or, as in the case of the addition of figures, the condition of unusually rapid performance of the work be imposed. If the interest be strongly excited, good results are often obtained, although the amount of absolute intellectual effort may be comparatively small. This is illustrated by the effect on some persons of the reading of interesting books, or of conversation of a like character.

But experience teaches that previous training of the subjects of experiment has a great deal to do with the results. Practice enables the individual experimented on to at once abstract the mind from everything except the work in hand. Moreover, a subject thus trained is careful that the mind be comparatively inactive for some time previous to the commencement of the

General Considerations.

experiment—a point of great importance—as, if it be neglected, a *fall* of temperature in the head may be obtained instead of a *rise*, if the degree of mental activity of the subject has been greater before than during the experiment. Such a subject occupied with work of no great intellectual difficulty may exhibit both more rapid and greater rises of temperature than can be obtained on untrained individuals, although the latter may be performing work requiring mental exertion of a far higher order.

It is often noticed that the same individual may be at different times very differently affected by the same kind of work, even when the mental effort appears to be about equal; the difference in the results seeming to be due to the varying degree of interest felt.

The first lesson to be taught the subject is *to stop thinking, at will.* Practice in this, and tact in the experimentor in diverting the mind of the subject from trains of thought, or in leading the mind into such trains, constitute a large part of the requirements of successful experimenting. Subjects soon learn to judge, with a certain degree of accuracy, from their own sensations, whether an experiment is succeeding or not. If the subject be interested in the result, it is often well to magnify the degree of success, or to conceal the existence of failure, encouragement thus given frequently stimulating the individual to fresh efforts. Struggling against obstacles, when the subject has become discouraged, is rarely productive of good results, whereas, as soon as a way out of the difficulty becomes evident, even though the mental strain be diminished, a marked effect on the temperature may be seen.

The kind of mental action which is reduced with most difficulty to experimental control is the emotions. We have here, in the majority of cases, to trust to chance; but an opportunity of studying systematically a portion of this subject offers itself in some persons who can arouse, almost voluntarily, certain mixed emotions by the reading or recitation of poetical or prose productions of an appropriate character. Of all the elevations of temperature due to mental activity which have been noted by the author, the greatest have been observed in this class of cases. Genuine emotion must, however, be experienced, as mere mechanical reading or recitation is of no effect. Moreover, recitation to one's self appears to have even a greater effect than recitation aloud.

With regard to other emotions, the effects of anger and vexation have been examined on several occasions, and also the effect of mirth in a few cases.

We proceed, in the next place, to consider the methods of experimenting.

In examining the influence of mental activity on the temperature of the head, considered without special reference to the *comparative* effect on different regions or sides, we have our choice of two methods of procedure. One method is to affix one pile to the head, and another pile with its current opposed to that of the first, to some other part of the body—the thigh, for example—thus obtaining the difference of temperature between the two parts. The other method is to apply one pile to the head, as before, and to use an opposing current of constant strength sufficient to keep the spot of light of the galvanometer on the scale when the pile upon the head is in action. In both methods it is well to employ a second galvanometer to test the strength of the opposing current. The temperature of the thigh often falls during prolonged mental exertion, sometimes slightly, at other times very markedly. This fall—as the author has noticed in former experiments already cited—is not due entirely to immobility of the limb, as this condition by itself, without concomitant mental exertion, does not produce so marked a depression of temperature as when the brain is active. One of the causes of the decrease of temperature would appear to be contraction of blood-vessels under vaso-motor influence, as Mosso has shown to occur in the forearm during increased cerebral action.*

If we wish to compare the temperatures of two spaces with each other, the best method is to apply a pile to each space, thus making a direct comparison.

The piles are secured to the head by means of tapes passing through eyes in metal rings, which surround the ebonite casing of the pile,† and are prevented from falling off by flanges in the casing near the face of the pile. The apparatus is the same as that already described in Part I.

The subject of the experiment should remain quietly seated,

---

* 'Archives de Physiologie,' 2nd série, t. iii, p. 177.

† See plate of author's apparatus in 'Archives de Physiologie,' July and August, 1868, vol. i.

General Considerations.

with the piles attached, for some minutes before commencing the observations, and the comfort of the individual with regard to the position of the head and body, temperature of the room, &c., must be carefully attended to at the start, so as to avoid future interruption. Moreover, the subject should be well instructed with reference to the danger to the success of the experiment of any movements of the head or body beyond certain fixed limits.

The piles must be well down upon the skin; and if the hair be thick, it must be thinned by cutting and shaving the inner layers, the outer layers being raised for this purpose.

We will now consider the effect of intellectual work on the temperature of different parts of the head. The following experiments were all made on the same individual, and under circumstances as nearly as possible identical. The particular spaces tested were chosen as being characteristic, as regards position, of the author's three arbitrary regions, and also because experience had shown that they gave good results, although, in this latter respect, it will be seen farther on that they are surpassed by other spaces.

In the experiments zero represents the temperature held by the space examined, at the commencement of the observations, and the deflections of the galvanometer, with the corresponding thermometric values, are the degrees of rise above this initial temperature. The deflections of the galvanometer are given as furnishing a more ready index to the eye of the changes of temperature than the thermometric values.

ANTERIOR REGION.

Four sets of experiments on 3rd district, 3rd tier, left side.

1st Set.

Mean of 10 observations. Mean temperature of air $12.77°$ C. ($55.98°$ F.). Mean temperature of space examined, at commencement of work, $33.7°$ C. ($92.66°$ F.). Work performed consisted of mathematical calculations of considerable difficulty performed rather slowly.

Temperature of the Head.

Rise of temperature.

Time from the commencement of work.	Deflections of galvanometer.	Thermometric values.
At the end of 5 minutes	3°	0·006° C. (0·0108° F.).
,, 10 ,,	5°	0·0100° C. (0·018° F.).
,, 25 ,,	8·2°	0·0164° C. (0·0295° F.).
,, 35 ,,	10°	0·020° C. (0·036° F.).
,, 45 ,,	12°	0·024° C. (0·043° F.).
,, 60 ,,	12·5°	0·025° C. (0·045° F.).
,, 75 ,,	14°	0·028° C. (0·05° F.).
,, 90 ,,	,,	,,

2nd Set.

Mean of 10 observations. Mean temperature of air 13° C. (55·4° F.). Mean temperature of space examined, at commencement of work, 34° C. (93·2° F.). Work performed consisted of arithmetical computations performed as rapidly as possible.

At the end of 5 minutes	4°	0·008° C. (0·0144° F.).
,, 10 ,,	7·5°	0·015° C. (0·027° F.).
,, 25 ,,	12°	0·024° C. (0·043° F.).
,, 35 ,,	13·1°	0·0262° C. (0·0471° F.).
,, 45 ,,	15·2°	0·0304° C. (0·054° F.).
,, 60 ,,	17°	0·034° C. (0·061° F.).
,, 75 ,,	,,	,,

3rd Set.

Mean of 10 observations. Mean temperature of air 12·5° C. (54·5° F.). Mean temperature of space examined, at commencement of work, 33·6° C. (92·48° F.). Work performed consisted in making notes of subjects requiring considerable reflection.

At the end of 5 minutes	0°	0°.
,, 10 ,,	3·3°	0·0086° C. (0·0154° F.).
,, 25 ,,	7·1°	0·0142° C. (0·0255° F.).
,, 35 ,,	9·2°	0·0184° C. (0·033° F.).
,, 45 ,,	10°	0·02° C. (0·036° F.).
,, 60 ,,	11°	0·022° C. (0·0396° F.).
,, 75 ,,	,,	,,
,, 90 ,,	,,	,,

General Effect of Intellectual Work.

4th Set.

Mean of 10 observations. Mean temperature of air 13° C. 55·4° F.). Mean temperature of space examined, at commencement of work, 34·2° C. (93·56° F.). Work performed consisted in putting into writing ideas which were difficult of expression.

Time from the commencement of work.	Deflections of galvanometer.	Rise of temperature. Thermometric values.
At the end of 5 minutes	. 5·2°	. 0·0104° C. (0·0187° F.).
,, 10 ,,	. 9°	. 0·018° C. (0·0324° F.).
,, 25 ,,	. 12·3°	. 0·0246° C. (0·044° F.).
,, 35 ,,	. 14°	. 0·028° C. (0·0504° F.).
,, 45 ,,	. 15·5°	. 0·031° C. (0·0558° F.).
,, 60 ,,	. 17·4°	. 0·0348° C. (0·0626° F.).
,, 75 ,,	. 18°	. 0·036° C. (0·0648° F.).
,, 90 ,,	. ,,	. ,, ,,

Rises of temperature in preceding 4 sets of experiments.

	Highest rises.	Average rises per minute in first thirty-five minutes.
1st set	0·028° C. (0·05° F.)	. 0·000571° C. (0·001027° F.).
2nd ,,	0·034° C. (0·061° F.)	. 0·000748° C. (0·001346° F.).
3rd ,,	0·022° C. (0·0396° F.)	. 0·000525° C. (0·000945° F.).
4th ,,	0·036° C. (0·0648° F.)	. 0·000800° C. (0·00144° F.).

Average highest rise in all four sets.	Average rises per minute in first thirty-five minutes in all four sets.
0·03° C. (0·054° F.).	0·000661° C. (0·001198° F.).

Middle Region.

Four sets of experiments on 3rd district, 5th tier, left side.

1st Set.

Mean of 10 observations. Mean temperature of air 13·75° C. (56·75° F.). Mean temperature of space examined, at commencement of work, 33·5° C. (92·3° F.). Work performed the same as in the corresponding set of experiments on anterior region.

Temperature of the Head.

Time from the commencement of work.	Deflections of galvanometer.	Rise of temperature.	
		Thermometric values.	
At the end of 5 minutes	3°	0·006° C.	(0·0108° F.).
,, 10 ,,	6°	0·012° C.	(0·0216° F.).
,, 25 ,,	7°	0·014° C.	(0·0252° F.).
,, 35 ,,	9·5°	0·019° C.	(0·0342° F.).
,, 45 ,,	11°	0·022° C.	(0·0396° F.).
,, 60 ,,	11·7°	0·0234° C.	(0·0421° F.).
,, 75 ,,	13·4°	0·0268° C.	(0·0482° F.).
,, 90 ,,	,,	,,	,,

2nd Set.

Mean of 10 observations. Mean temperature of air 13° C. (55·4° F.). Mean temperature of space examined, at commencement of work, 33·6° C. (92·48° F.). Work performed the same as in the corresponding set of experiments on anterior region.

At the end of 5 minutes	4·2°	0·0084° C.	(0·0151° F.).
,, 10 ,,	7·3°	0·0146° C.	(0·0262° F.).
,, 25 ,,	10°	0·020° C.	(0·036° F.).
,, 35 ,,	10·4°	0·0208° C.	(0·0374° F.).
,, 45 ,,	14°	0·028° C.	(0·0504° F.).
,, 60 ,,	15°	0·030° C.	(0·054° F.).
,, 75 ,,	16·3°	0·0326° C.	(0·0586° F.).
,, 90 ,,	,,	,,	,,

3rd Set.

Mean of 10 observations. Mean temperature of air 14° C. (57·2° F.) Mean temperature of space examined, at commencement of work, 33·4° C. (92·12° F.). Work performed the same as in the corresponding set of experiments on anterior region.

At the end of 5 minutes	0°	0°.	
,, 10 ,,	4°	0·008° C.	(0·0144° F.).
,, 25 ,,	6°	0·012° C.	(0·0216° F.).
,, 35 ,,	7·75°	0·0155° C.	(0·0279° F.).
,, 45 ,,	8·5°	0·017° C.	(0·0306° F.).
,, 60 ,,	9°	0·018° C.	(0·0324° F.)
,, 75 ,,	9·75	0·0195 C.	(0·0351° F.).
,, 90 ,,	,,	,,	,,

General Effect of Intellectual Work.

4th Set.

Mean of 10 observations. Mean temperature of air 13·4° C. (56·12° F.). Mean temperature of space examined, at commencement of work, 33·5° C. (92·3° F.). Work performed the same as in the corresponding set of experiments on anterior region.

Rise of temperature.

Time from the commencement of work.	Deflections of galvanometer.	Thermometric values.
At the end of 5 minutes	. 6°	. 0·012° C. (0·0216° F.).
,, 10 ,,	. 8·5°	. 0·017° C. (0·0306° F.).
,, 25 ,,	. 11·8°	. 0·0236° C. (0·0424° F.).
,, 35 ,,	. 13°	. 0·026° C. (0·0468° F.).
,, 45 ,,	. 14·5°	. 0·029° C. (0·0522° F.).
,, 60 ,,	. 15·75°	. 0·0315° C. (0·0567° F.).
,, 75 ,,	. 16·5°	. 0·033° C. (0·0594° F.).
,, 90 ,,	. 17·5°	. 0·035° C. (0·063° F.).
,, 100 ,,	. ,,	. ,, ,,

Rises of temperature in preceding 4 sets of experiments.

	Highest rises.	Average rises per minute in first thirty-five minutes.
1st set	0·0268° C. (0·048° F.).	. 0·000542° C. (0·000975° F.).
2nd ,,	0·0326° C. (0·0568° F.).	. 0·000594° C. (0·001069° F.).
3rd ,,	0·0195° C. (0·0351° F.).	. 0·000442° C. (0·000795° F.).
4th ,,	0·0350° C. (0·063° F.).	. 0·000742° C. (0·001335° F.).

Average highest rise in all four sets.	Average rise per minute in first thirty-five minutes in all four sets.
0·0284° C. (0·0511° F.).	0·00058° C. (0·001044° F.).

Posterior Region.

Four sets of experiments on 2nd district, 4th tier, left side.

1st Set.

Mean of 10 observations. Mean temperature of air 12·5° C. (54·5° F.). Mean temperature of space examined, at commencement of work, 33·6° C. (92·48° F.). Work performed the same as in the corresponding set of experiments on anterior region.

130   *Temperature of the Head.*

	Time from the commencement of work	Deflections of galvanometer.	Rise of temperature. Thermometric values.
At the end of	5 minutes	2°	0·004° C. (0·0072° F.).
,,	10 ,,	4°	0·008° C. (0·0144° F.).
,,	25 ,,	4·5°	0·009° C. (0·0342° F.).
,,	35 ,,	8·5°	0·0170° C. (0·0306° F.).
,,	45 ,,	9°	0·018° C. (0·0324° F.).
,,	60 ,,	10°	0·020° C. (0·036° F.).
,,	75 ,,	,,	,,

2nd Set.

Mean of 10 observations. Mean temperature of air 12° C. (53·6° F.). Mean temperature of space examined, at commencement of work, 33·4° C. (92·12° F.). Work performed the same as in the corresponding set of experiments on anterior region.

At the end of	5 minutes	3°	0·006° C. (0·0108° F.).
,,	10 ,,	5°	0·010° C. (0·018° F.).
,,	25 ,,	7·5°	0·015° C. (0·027° F.).
,,	35 ,,	9·2°	0·0184° C. (0·0331° F.).
,,	45 ,,	11°	0·022° C. (0·0396° F.).
,,	60 ,,	12°	0·024° C. (0·0432° F.).
,,	75 ,,	14·1°	0·0282° C. (0·0507° F.).
,,	90 ,,	,,	,,

3rd Set.

Mean of 10 observations. Mean temperature of air 13° C. (55·4° F.). Mean temperature of space examined, at commencement of work, 33·5° C. (92·3° F.). Work performed the same as in the corresponding set of experiments on anterior region.

At the end of	5 minutes	0°	0°.
,,	10 ,,	2°	0·004° C. (0·0072° F.).
,,	25 ,,	4·2°	0·0084° C. (0·0151° F.).
,,	35 ,,	5·5°	0·011° C. (0·0198° F.).
,,	45 ,,	7°	0·014° C. (0·0252° F.).
,,	60 ,,	8·7°	0·0174° C. (0·0313° F.).
,,	75 ,,	8·9°	0·0178° C. (0·032° F.).
,,	90 ,,	,,	,,

General Effect of Intellectual Work.

4th Set.

Mean of 10 observations. Mean temperature of air 13·5° C. (56·3° F.). Mean temperature of space examined, at commencement of work, 33·6° C. (92·48° F.). Work performed the same as in the corresponding set of experiments on anterior region.

Time from the commencement of work.	Rise of temperature.	
	Deflections of galvanometer.	Thermometric values.
At the end of 5 minutes	4°	0·008° C. (0·0144° F.).
,, 10 ,,	7°	0·014° C. (0·0252° F.).
,, 25 ,,	9°	0·018° C. (0·0324° F.).
,, 35 ,,	10·2°	0·0204° C. (0·0367° F.).
,, 45 ,,	11·5°	0·023° C. (0·0414° F.).
,, 60 ,,	12·75°	0·0255° C. (0·0459° F.).
,, 75 ,,	14·5°	0·029° C. (0·0522° F.).
,, 90 ,,	,,	,, ,,

Rises of temperature in preceding 4 sets of experiments.

	Highest rises.	Average rises per minute in first thirty-five minutes.
1st set	0·020° C. (0·036° F.)	0·000485° C. (0·000873° F.).
2nd ,,	0·0282° C. (0·0507° F.)	0·000525° C. (0·000945° F.).
3rd ,,	0·0178° C. (0·032° F.)	0·000314° C. (0·000565° F.).
4th ,,	0·029° C. (0·0522° F.)	0·000582° C. (0·001047° F.).

Average highest rise in all four sets.	Average rise per minute in first thirty-five minutes in all four sets.
0·0237° C. (0·0426° F.)	0·000477° C. (0·000858° F.).

COMPARISON OF THE THREE REGIONS WITH EACH OTHER.

Comparison of highest rises and average rates of rise.

	Average highest rises.	Average rises per minute in first thirty-five minutes.
Anterior region	0·03° C. (0·054° F.)	0·000661° C. (0·001189° F.).
Middle ,,	0·0284° C. (0·0511° F.)	0·000580° C. (0·001044° F.)
Posterior ,,	0·0237° C. (0·0426° F.)	0·000477° C. (0·000858° F.)

Temperature of the Head.

Average highest rise in all three regions taken together.
0·0273° C. (0·0491° F.)

Average rise per minute in first thirty-five minutes in all three regions taken together.
0·000572° C. (0·001029° F.).

Comparison of effects of different kinds of work on each region.

	Highest rises.			Average rise in each set of experiments for the three regions taken together.
	Anterior.	Middle.	Posterior.	
1st set	0·028° C. (0·0504° F.).	0·0268° C. (0·0482 F.).	0·0200° C. (0·036° F.).	0·0249° C. (0·0448° F.).
2nd „	0·034° C. (0·0612° F.).	0·0326° C. (0·0586° F.).	0·0282° C. (0·0507° F.).	0·0316° C. (0·0568° F.).
3rd „	0·022° C. (0·0396° F.).	0·0195° C. (0·0351° F.).	0·0178° C. (0·032° F.).	0·0197° C. (0·0354° F.).
4th „	0·036° C. (0·0648° F.).	0·035° C. (0·063° F.).	0·029° C. (0·0522° F.).	0·0333° C. (0·0599° F.).

The preceding results indicate :

1st. That intellectual work of the four varieties enumerated causes a rise of temperature in all three regions of the head.

2nd. That the rapidity and the degree of this rise are different in the different regions.

3rd. That different kinds of work produce, in the same region, elevations of temperature differing both in degree and in rapidity of appearance.

4th. That the kind of work which affects one region in the greatest or least degree affects similarly the other regions.

5th. That, in the particular spaces examined, the order of the regions with regard to both degree and rapidity of rise of temperature, in all four kinds of work specified, is :—1st. Anterior region ; 2nd. Middle region ; 3rd. Posterior region.

6th. That, in the particular individual experimented on, the greatest effect, both as regards degree and rapidity of rise, was produced by the work of the 4th set of experiments (composition) ; the next greatest effect was produced by the work of the 2nd set (rapid arithmetical computations) ; and the least effect was produced by the work of the 3rd set (making of notes).

Of the above conclusions, the last two require, under altered conditions regarding the particular spaces examined, and the in-

General Effect of Intellectual Work. 133

dividuals experimented on, important modifications. It will be seen, farther on, that the anterior region by no means invariably shows the highest rise of temperature, but that, on the contrary, certain points of the middle region surpass it in this respect. With regard to individual peculiarities, enough has already been said to show that no rule can be laid down as to the comparative effect of the same kind of work in different persons, or even in the same person at different times; neither the intrinsic difficulty of the work itself, nor the special difficulty it may offer to the individual, being the sole factors to be taken into account, but the degree of interest felt by the worker being also involved.

CHAPTER II.

EXPERIMENTS ILLUSTRATIVE OF THE EFFECT OF INTELLECTUAL WORK ON THE TEMPERATURE OF DIFFERENT PARTS OF THE HEAD.—AVERAGE RISES OF TEMPERATURE FOR THE DIFFERENT REGIONS, IN INTELLECTUAL WORK.—RISES OF TEMPERATURE ABOVE THE AVERAGE, IN INTELLECTUAL WORK.

BUT experiments of the kind given in the last chapter represent comparatively favorable cases, where the course of the observations was uninterrupted, and where, therefore, a gradual, steady rise of temperature was maintained pretty thoroughly from the commencement. Such cases are not, however, the rule, and are given here principally to simplify the demonstration of the important fundamental facts contained in the first four conclusions just stated. More frequently we have to deal with less regular variations of temperature.

The experiments to which attention will now be called have been selected from a large number made on different spaces of the several regions on different subjects, and are intended to give an idea of the uncertain fluctuations of temperature commonly met with, and of the relation of these fluctuations to coincident variations of the mental condition. It will likewise be observed, in some of these experiments, that the mental work is several times purposely interrupted, and also that the nature of the work is altered. The times of observation are also more numerous than in the experiments just given. In the experiments the "plus" sign indicates a rise above, and the "minus" sign a fall below the original temperature, at the commencement of the observations, of the space examined, this starting-point being represented, as in the preceding experiments, by zero. Where words indicating the cessation or the commencement of work are placed, in the course of an experiment, opposite to the space between two observations, they always refer to the time *immediately preceding* them. Thus the words, "Stopped work," placed opposite to the space between the observations of the fourteenth and seventeenth minutes, in 1st experiment, anterior region, signify that the work was stopped *at the end of the fourteenth minute.* Other words expressive of the mental condition apply to the times *directly opposite to which they are placed.*

The results of the different experiments are not comparable with each other, as they have been obtained on different individuals, at different times, and under dissimilar circumstances. Two examples are given from each region.

Anterior Region.

1st *Experiment.*

Examination of 4th district, 2nd—3rd tier, left side.

Time from commencement of work.	Rise or fall of temperature.		Mental condition.
	Deflections of galvanometer.	Thermometric values.	
At the end of—			
0 minutes	$0°$	$0°$	Commenced mathematical calculations.
3 „	$+6°$	$+0·012°$ C. ($0·0216°$ F.)	
10 „	$+3°$	$+0·006°$ C. ($0·0108°$ F.)	} Interest flagging.
11 „	$+4°$	$+0·008°$ C. ($0·0144°$ F.)	
12 „	$+6°$	$+0·012°$ C. ($0·0216°$ F.)	Interest regained.
14 „	$+11°$	$+0·022°$ C. ($0·0396°$ F.)	
			Stopped work.
17 „	$+7°$	$+0·014°$ C. ($0·0252°$ F.)	
			Commenced making notes requiring reflection.
19 „	$+5°$	$+0·010°$ C. ($0·018°$ F.)	
$19\frac{1}{2}$ „	$+7°$	$+0·014°$ C. ($0·0252°$ F.)	
21 „	$+11°$	$+0·022°$ C. ($0·0396°$ F.)	
			Stopped work.
22 „	$+9°$	$+0·018°$ C. ($0·0324°$ F.)	} Thinking more or less actively.
23 „	$+8·5°$	$+0·017°$ C. ($0·0306°$ F.)	
$23\frac{1}{2}$ „	$+9°$	$+0·018°$ C. ($0·0324°$ F.)	
24 „	$+8°$	$+0·016°$ C. ($0·0288°$ F.)	
			Ceased thinking to any extent.
26 „	$+5°$	$+0·010°$ C. ($0·018°$ F.)	
27 „	$+3°$	$+0·006°$ C. ($0·0108°$ F.)	

In the above experiment the rise of temperature due to the mathematical work ceased after the third minute, and at the tenth and eleventh minutes a return toward the starting point had been effected. Coincident with this thermal decrease was a partial loss of interest in the work. At the twelfth minute the interest had

Temperature of the Head.

been regained, and the temperature had also regained the point held at the third minute. The temperature then rose until the work was stopped, at the end of the fourteenth minute. After an interval of three minutes, during which the temperature fell 0·008° C. (0·0144° F.), work of a different kind was begun, and at the expiration of four minutes, when the work ended, the temperature had risen again to the point occupied at the close of the mathematical work. The temperature after this sank toward the starting point, the rapidity of its fall being, however, somewhat impeded by the state of activity of the mind during part of the time (22nd to 23½rd minutes).

2nd Experiment.
Examination of 3rd district, 2nd—3rd tier, right side.

Time from commencement of work.	Rise or fall of temperature.		Mental condition.
	Deflections of galvanometer.	Thermometric values.	
At the end of—			
0 minutes	0°	0°	Commenced mathematical calculations.
4½ ,,	+4°	+0·008° C. (0·0144° F.)	
9 ,,	+4·5°	+0·009° C. (0·0162° F.)	
10 ,,	+ ,,	+ ,, ,,	Stopped work, and commenced conversation; not much interested.
12 ,,	+3°	+0·006° C. (0·0108° F.)	
15 ,,	0°	0°	Stopped talking, and commenced reading aloud matter of little or no interest.
18 ,,	−3°	−0·006° C. (0·0108° F.)	
24½ ,,	+6°	+0·012° C. (0·0216° F.)	Stopped reading, and commenced mathematics again.
25 ,,	+6·5°	+0·013° C. (0·0234° F.)	
27½ ,,	+5°	+0·010° C. (0·018° F.)	Stopped work.
29 ,,	+4°	+0·008° C. (0·0144° F.)	
31 ,,	+2·5°	+0·005° C. (0·009° F.)	
33 ,,	+1·5°	+0·003° C. (0·0054° F.)	
41 ,,	+9°	+0·018° C. (0·0324° F.)	Commenced reading to one's self; much interested.
42 ,,	+12°	+0·024° C. (0·0432° F.)	
43 ,,	+13°	+0·026° C. (0·0468° F.)	
45 ,,	+3°	+0·006° C. (0·0108° F.)	Stopped reading. Commenced mathematics again.
47½ ,,	+6·5°	+0·013° C. (0·0234° F.)	
49 ,,	+10·5°	+0·021° C. (0·0378° F.)	

General Effect of Intellectual Work.

In this experiment mathematical work caused, in the first nine minutes, a rise of temperature of 0·009° C. (0·0162° F.), which was succeeded by a fall of 0·003° C. (0·0054° F.), during the three minutes of conversation immediately following the cessation of work. From the end of the twelfth to the end of the eighteenth minute the subject was occupied in reading aloud matter of an uninteresting character. During this time, the temperature fell still further, reaching a point, 0·006° C. (0·0108° F.), below the original starting point. The resumption of mathematical work caused a rise of 0·019° C. (0·0342° F.), (0·012° C.—0·0216° F., above the original starting point) in seven minutes. Work being now stopped again, the temperature fell to 0·003° C. (0·0054° F.), when reading to one's self of very interesting matter was begun. The result of the reading was a rise in ten minutes of 0·023° C. (0·0414° F.). At the end of this time the reading was discontinued, and a fall of temperature ensued of 0·02° C. (0·036° F.) in two minutes; which fall was succeeded by another rise of 0·015° C. (0·027° F.) in four minutes, due to mathematical work.

MIDDLE REGION.

1st Experiment.

Examination of 1st district, 4th tier, left side.

Time from commencement of work.	Rise or fall of temperature.		Mental condition.
	Deflections of galvanometer.	Thermometric values.	
At the end of—			
0 minutes	0°	0°	Commenced reading aloud matter of interest.
1 ,,	+1°	+0·002° C. (0·0036° F.)	
2 ,,	+2°	+0·004° C. (0·0072° F.)	
			Reading interrupted.
3¾ ,,	−1°	−0·002° C. (0·0036° F.)	
5½ ,,	− ,,	− ,, ,,	
			Reading resumed.
6½ ,,	+5°	+0·010° C. (0·018° F.)	⎤
7¾ ,,	+3°	+0·006° C. (0·0108° F.)	⎥
9 ,,	+ ,,	+ ,, ,,	⎥
9¾ ,,	+6·5°	+0·013° C. (0·0234° F.)	⎬ Degree of interest variable.
11 ,,	+9°	+0·018° C. (0·0324° F.)	⎥
11½ ,,	+8°	+0·016° C. (0·0288° F.)	⎥
12 ,,	+7°	+0·014° C. (0·0252° F.)	⎦

Temperature of the Head.

Time from commencement of work.	Rise or fall of temperature.		Mental condition.
	Deflections of galvanometer.	Thermometric values.	

At the end of—

13 minutes	. +8°	. +0·016° C.	(0·0288° F.)	⎫
14 ,,	. +13°	. +0·026° C.	(0·0468° F.)	⎪
15½ ,,	. +10°	. +0·02°C.	(0·036° F.)	⎪
16½ ,,	. +7°	. +0·014° C.	(0·0252 F.)	⎬ Degree of interest variable.
17 ,,	. +8°	. +0·016° C.	(0·0288° F.)	⎪
17½ ,,	. +6°	. +0·012° C.	(0·0216° F.)	⎪
19 ,,	. +8°	. +0·016° C.	(0·0288° F.)	⎪
19½ ,,	. +4·5°	. +0·009° C.	(0·0162° F.)	⎭
20 ,,	. +8°	. +0·016° C.	(0·0288° F.)	⎫
20½ ,,	. +9°	. +0·018° C.	(0·0324° F.)	⎪
21½ ,,	. +13°	. +0·026° C.	(0·0468° F.)	⎬ Interest steadily increasing.
22½ ,,	. +16°	. +0·032° C.	(0·0576° F.)	⎪
23½ ,,	. +19°	. +0·038° C.	(0·0684° F.)	⎭
				Stopped reading.
33 ,,	. +5°	. +0·010° C.	(0·018° F.)	

In the above experiment attention is called to the number and extent of the fluctuations of temperature occurring during the eighteen minutes of the last reading, the degree of interest experienced by the reader during this time being very variable.

2nd Experiment.

Examination of 1st district, 3rd tier, right side.

At the end of—

0 minutes	. 0°	. 0°		Commenced mathematical calculations.
2½ ,,	. +3·5°	. +0·007° C.	(0·0126° F.)	
7 ,,	. +4°	. +0·008° C.	(0·0144° F.)	
12 ,,	. +5·5°	. +0·011° C.	(0·0198° F.)	
14½ ,,	. +10·5°	+0·021° C.	(0·0378° F.)	
18½ ,,	. +11·5°	+0·023° C.	(0·0414° F.)	
22 ,,	·. +9·5°	. +0·019° C.	(0·0342° F.)	
24½ ,,	. +7·5°	. +0·015° C.	(0·027° F.)	⎫
27 ,,	. +6·5°	. +0·013° C.	(0·0234° F.)	⎬ Interest flagging somewhat.
31 ,,	. +4·5°	. +0·009° C.	(0·0162° F.)	⎭
33½ ,,	. +5·5°	. +0·011° C.	(0·0198° F.)	
46½ ,,	. +7·5°	. +0·015° C.	(0·027° F.)	
50½ ,,	. +8°	. +0·016° C.	(0·0288° F.)	
				Stopped mathematical work, and commenced making notes requiring reflection.
74 ,,	. +12°	. +0·024° C.	(0·0432° F.)	

General Effect of Intellectual Work. 139

	Rise or fall of temperature.		
Time from commencement of work.	Deflections of galvanometer.	Thermometric values.	Mental condition.

At the end of—

75½ minutes	. +13·5°	+0·027° C. (0·0486° F.)	
			Stopped work.
80½ ,,	. +9·5°.	+0·019° C. (0·0342° F.)	
85½ ,,	. +3·5°.	+0·007° C. (0·0126° F.)	

In this experiment, after a steady rise of 0·023° C. (0·0414° F.) in the first eighteen and a half minutes of mathematical work, the temperature steadily declined to 0·009° C. (0·0162° F.) above the starting point, this last value being reached twelve and a half minutes from the cessation of the rise. During the greater part of this time the subject of the experiment was quite conscious of partial loss of interest in the work. In the ensuing forty-four and a half minutes the temperature rose to 0·027°C. (0·0486°F.), this being above its first maximum, the nature of the work being, moreover, changed during a part of this time.

POSTERIOR REGION.

1st Experiment.

Examination of 1st district, 3rd—4th tier, left side.

At the end of—

0 minutes	. 0°	. 0°	Commenced mathematical
7 ,,	. +4°	. +0·008° C. (0·0144° F.)	calculations.
9 ,,	. +6°	. +0·012° C. (0·0216° F.)	
11⅓ ,,	. +7·5°.	+0·015° C. (0·027° F.)	
14½ ,,	. +9°	. +0·018° C. (0·0324° F.)	
17½ ,,	. +10°	. +0·02° C. (0·036° F.)	
22 ,,	. + ,,	. + ,,	,,
26½ ,,	. +8°	. +0·016° C. (0·0288° F.)	
30 ,,	. +5°	. +0·01° C. (0·018° F.)	⎫
34 ,,	. +5·5°.	+0·011° C. (0·0198° F.)	⎬ Interest decidedly flagging.
35 ,,	. +4°	. +0·008° C. (0·0144° F.)	⎭
35½ ,,	. + ,,	. + ,,	,,
39 ,,	. +1°	. +0·002° C. (0·0036° F.)	
41½ ,,	. +6°	. +0·012° C. (0·0216° F.)	Interest regained.
43½ ,,	. + ,,	. + ,,	,,
52 ,,	. +9°	. +0·018° C. (0·0324° F.)	
54 ,,	. + ,,	. + ,,	,,

Temperature of the Head.

Time from commencement of work.	Rise or fall of temperature.		Mental condition.
	Deflections of galvanometer.	Thermometric values.	

At the end of—

56 minutes	+5°	+0·01° C.	(0·018° F.)	
60 ,,	+1°	+0·002° C.	(0·0036° F.)	} Interest again failing.
65 ,,	+ ,,	+ ,,	,,	
68½ ,,	+4·5°	+0·009° C.	(0·0162° F.)	Interest returning.
71 ,,	+7°	+0·014° C.	(0·0252° F.)	
74 ,,	+10·5°	+0·021° C.	(0·0378° F.)	
76½ ,,	+11°	+0·022° C.	(0·0396° F.)	
				Stopped work.
77½ ,,	+8°	+0·016° C.	(0·0288° F.)	
78 ,,	+7°	+0·014° C.	(0·0252° F.)	
79 ,,	+6°	+0·012° C.	(0·0216° F.)	
80 ,,	+4°	+0·008° C.	(0·0144° F.)	

This experiment shows, at two periods, a cessation of rise, and a fall of temperature nearly to the original starting-point; both periods being characterised by a failure of interest in the work being done.

2nd Experiment.

Examination of 3rd district, 3rd tier, left side.

At the end of—

0 minutes	0°	0°		Commenced reading aloud.
1 ,,	+1°	+0·002° C.	(0·0036° F.)	
3½ ,,	−1°	−0·002° C.	(0·0036° F)	⎫
5 ,,	+1°	+0·002° C.	(0·0036° F.)	
6½ ,,	+2°	+0·004° C.	(0·0072° F.)	
7½ ,,	+5°	+0·01° C.	(0·018° F.)	
8¾ ,,	+3°	+0·006° C.	(0·0108° F.)	
9 ,,	+6°	+0·012° C.	(0·0216° F.)	Decided lack of interest
11 ,,	+1°	+0·002° C.	(0·0036° F.)	⎬ and attention at times.
12 ,,	+3°	+0·006° C.	(0·0108° F.)	
15 ,,	+5°	+0·01° C.	(0·018° F.)	
17 ,,	+3°	+0·006° C.	(0·0108° F.)	
19½ ,,	+ ,,	+ ,,	,,	
20 ,,	+4°	+0·008° C.	(0·0144° F.)	
26½ ,,	+2·5°	+0·005° C.	(0·009° F.)	⎭
27½ ,,	+5°	+0·010° C.	(0·018° F.)	
28 ,,	+6°	+0·012° C.	(0·0216° F.)	
30 ,,	+7°	+0·014° C.	(0·0252° F.)	

During the whole of the above experiment the temperature

was unsteady, rising and falling with great irregularity. The reader was at times inattentive to the work, feeling no interest in the matter read, and reading almost mechanically. At other times the reader became momentarily interested, and read with attention.

Nearly every one of the 88 spaces on a side, into which the surface of the head has been divided in these investigations, has been examined with reference to the effect upon its temperature of increased intellectual activity of different kinds; and it may be said that, so far as the author's experience goes, every space is capable of showing an elevation of temperature in every kind of intellectual work. Certain spaces, however, exhibit, as a rule, higher rises of temperature than others, as will be hereafter seen.

The following spaces in the three regions have been specially examined; and the average rises of temperature, due to intellectual work of all kinds, in the spaces belonging to each region, are as follows:

Anterior region.

Spaces examined.	Average rise of temperature for all the spaces of each region.
2nd district, 2nd tier	
3rd ,, 2nd ,,	
3rd ,, 3rd ,,	
3rd–4th ,, 5th ,,	$0.034°$ C. ($0.0612°$ F.).
4th ,, 2nd–3rd ,,	
5th ,, 1st–2nd ,,	
5th ,, 3rd ,,	

Middle region.

Spaces examined.	Average rise of temperature.
1st district, 2nd tier	
1st ,, 3rd ,,	
1st–2nd ,, 4th ,,	
2nd ,, 3rd ,,	
2nd ,, 5th ,,	
3rd ,, 2nd–3rd ,,	
3rd ,, 3rd ,,	$0.0375°$ C. ($0.0675°$ F.).
3rd ,, 4th ,,	
3rd ,, 5th ,,	
4th ,, 2nd–3rd ,,	
4th ,, 4th ,,	
4th ,, 6th ,,	
4th ,, 7th ,,	

Temperature of the Head.

Posterior region.

Spaces examined.			Average rise of temperature for all the spaces of each region.
1st district,	3rd	tier	
2nd "	4th	"	
2nd "	6th	"	0·0296° C. (0·0532° F.).
3rd "	2nd	"	
3rd "	3rd	"	
5th "	3rd	"	

Average for all three regions, 0·0337° C. (0·0606° F.).

But much higher rises of temperature are not unfrequently seen in each region than the averages just given. The following are specimens of these high rises of temperature:

ANTERIOR REGION.

Examination of 5th district, 3rd tier, left side.

Time from commencement of work.	Rise or fall of temperature.		Mental condition.
	Deflections of galvanometer.	Thermometric values.	
At the end of—			
0 minutes	0°	0°	Commenced mathematical calculations.
1 "	+1°	+0·002° C. (0·0036° F.)	
4 "	+15·5°	+0·031° C. (0·0558° F.)	
5 "	+16·5°	+0·033° C. (0·0594° F.)	
8 "	+28°	+0·056° C. (0·1008° F.)	
16 "	+26°	+0·052° C. (0·0936° F.)	
20 "	+24·5°	+0·049° C. (0·088° F.)	
24 "	+36·5°	+0·073° C. (0·1314° F.)	
26 "	+40·5°	+0·081° C. (0·1458° F.)	
29 "	+42·5°	+0·085° C. (0·153° F.)	
33 "	+41°	+0·082° C. (0·1476° F.)	
34 "	+36·5°	+0·073° C. (0·1314° F.)	
			Stopped work.
36 "	+24·5°	+0·049° C. (0·088° F.)	
36½ "	+20·5°	+0·041° C. (0·0738° F.)	
			Commenced making notes requiring thought.
39 "	+37·5°	+0·075° C. (0·135° F.)	
40 "	+39·5°	+0·079° C. (0·0142° F.)	
			Stopped work.
50 "	+12·5°	+0·025° C. (0·045° F.)	

General Effect of Intellectual Work. 143

Time from commencement of work.	Rise or fall of temperature.		Mental condition.
	Deflections of galvanometer.	Thermometric values.	

At the end of—
52½ minutes	. +18·5°	+0·037° C. (0·066° F.)	
56 ,,	. +12·5°	+0·025° C. (0·045° F.)	
			Commenced mathematics again.
61 ,,	. +24·5°	+0·049° C. (0·088° F.)	
61½ ,,	. +26·5°	+0·053° C. (0·0954 F.)	
62 ,,	. +29°	. +0·058° C. (0·104° F.)	
			Stopped work.
64½ ,,	. +22°	. +0·044° C. (0·0792° F.)	
67 ,,	. +15°	. +0·03° C. (0·054° F.)	

In this experiment the temperature rose 0·085° C. (0·153° F.) above its original starting-point, in the first twenty-nine minutes of the observations.

Middle Region.

Examination of 1st—2nd district, 4th tier, left side.

At the end of—
0 minutes	. 0°	. 0°	Commenced mathematical calculations.
5 ,,	. +11°	. +0·022° C. (0·0396° F.)	
6½ ,,	. +14°	. +0·028° C. (0·0504° F.)	
12 ,,	. +27°	. +0·054° C. (0·0972° F.)	
13 ,,	. +33°	. +0·066° C. (0·1188° F.)	
17 ,,	. +31°	. +0·062° C. (0·1116° F.)	
			Stopped work.
19 ,,	. +30°	. +0·06° C. (0·108° F.)	
			Commenced reading to one's self matter of
27 ,,	. +44°	. +0·088° C. (0·1584° F.)	
32 ,,	. +46°	. +0·092° C. (0·1656° F.)	much interest.
			Stopped reading.
34 ,,	. +44°	. +0·088° C. (0·1584° F.)	
38 ,,	. +36·5°	+0·073° C. (0·1314° F.)	
42 ,,	. +34°	. +0·068° C. (0·1224° F.)	
43½ ,,	. +31°	. +0·062° C. (0·1116° F.)	
47 ,,	. +28°	. +0·056° C. (0·1008° F.)	
50½ ,,	. + ,,	. + ,, ,,	
51 ,,	. +25°	. +0·05° C. (0·09° F.)	
53 ,,	. +23°	. +0·046° C. (0·0828° F.)	
54 ,,	. +20°	. +0·04° C. (0·072° F.)	
55 ,,	. +18°	. +0·036° C. (0·0648° F.)	

Temperature of the Head.

Time from commence- ment of work.	Rise or fall of temperature.		Mental condition.
	Deflections of galvanometer.	Thermometric values.	

At the end of—

57 minutes	. $+17°$	. $+0\cdot034°$ C. ($0\cdot0612°$ F.)	
59 ,,	. $+19°$	. $+0\cdot038°$ C. ($0\cdot0684°$ F.)	Commenced mathematics again.
63 ,,	. $+21°$	. $+0\cdot042°$ C. ($0\cdot0756°$ F.)	
67 ,,	. $+$,,	. $+$,, ,,	
			Stopped work.
70 ,,	. $+12°$	. $+0\cdot024°$ C. ($0\cdot0432°$ F.)	
71 ,,	. $+10°$	. $+0\cdot02°$ C. ($0\cdot036°$ F.)	
72 ,,	. $+8°$	. $+0\cdot016°$ C. ($0\cdot0288°$ F.)	
73 ,,	. $+4°$	. $+0\cdot008°$ C. ($0\cdot0144°$ F.)	
74 ,,	. $+3°$	. $+0\cdot006°$ C. ($0\cdot0108°$ F.)	

In the above experiment the rise of temperature above the original starting-point was $0\cdot092°$ C. ($0\cdot1656°$ F.), the result of mathematical work and reading combined. This rise was effected in the first thirty-two minutes of the experiment; two of these minutes were, however, occupied by the interval of inactivity between the cessation of the mathematical work and the commencement of the reading. The time of the rise of temperature was, therefore, more properly thirty minutes.

Posterior Region.

Examination of 5th district, 3rd tier, left side.

At the end of—

0 minutes	. $0°$	. $0°$	Commenced mathematical work.
4 ,,	. $+2°$	. $+0\cdot004°$ C. ($0\cdot0072°$ F.)	
9 ,,	. $+5°$	. $+0\cdot01°$ C. ($0\cdot018°$ F.)	
12 ,,	. $+9°$	. $+0\cdot018°$ C. ($0\cdot0324°$ F.)	
$13\frac{1}{2}$,,	. $+8°$	. $+0\cdot016°$ C. ($0\cdot0288°$ F.)	
18 ,,	. $+$,,	. $+$,, ,,	
22 ,,	. $+9\cdot5°$	. $+0\cdot019°$ C. ($0\cdot0342°$ F.)	
29 ,,	. $+12°$	. $+0\cdot024°$ C. ($0\cdot0432°$ F.)	
$29\frac{1}{2}$,,	. $+14°$	. $+0\cdot028°$ C. ($0\cdot0504°$ F.)	
$35\frac{1}{2}$,,	. $+19°$	. $+0\cdot038°$ C. ($0\cdot0684°$ F.)	
$45\frac{1}{2}$,,	. $+21°$	. $+0\cdot042°$ C. ($0\cdot0756°$ F.)	
46 ,,	. $+22°$	. $+0\cdot044°$ C. ($0\cdot0792°$ F.)	
			Stopped work.
47 ,,	. $+21\cdot5°$	$+0\cdot043°$ C. ($0\cdot0864°$ F.)	

General Effect of Intellectual Work. 145

Time from commencement of work.	Rise or fall of temperature.		Mental condition.
	Deflections of galvanometer.	Thermometric values.	

At the end of—

48½ minutes	$+21\cdot5°$	$+0\cdot043°$ C. ($0\cdot0864°$ F.)		
49 ,,	$+21°$	$+0\cdot042°$ C. ($0\cdot0756°$ F.)	⎫	
50 ,,	$+22°$	$+0\cdot044°$ C. ($0\cdot0792°$ F.)	⎪	
53 ,,	$+23°$	$+0\cdot046°$ C. ($0\cdot0828°$ F.)	⎬	Interested during this time in trying to overhear a conversation in an adjoining room.
55 ,,	$+24°$	$+0\cdot048°$ C. ($0\cdot0864°$ F.)	⎪	
56 ,,	$+23°$	$+0\cdot046°$ C. ($0\cdot0828°$ F.)	⎪	
57½ ,,	$+$,,	$+$,, ,,	⎪	
68 ,,	$+24°$	$+0\cdot048°$ C. ($0\cdot0864°$ F.)	⎭	
70 ,,	$+20°$	$+0\cdot04°$ C. ($0\cdot072°$ F.)		
71 ,,	$+19°$	$+0\cdot038°$ C. ($0\cdot0684°$ F.)		
72 ,,	$+$,,	$+$,, ,,		
73 ,,	$+17\cdot5°$	$+0\cdot035°$ C. ($0\cdot063°$ F.)		

In this experiment mathematical work caused a rise of temperature of $0\cdot044°$ C. ($0\cdot0792°$ F.) in forty-six minutes, when the work ceased. Three minutes later—the temperature having meanwhile fallen $0\cdot002°$ C. ($0\cdot0036°$ F.)—the subject of the experiment becoming much interested in listening to a conversation near at hand, the temperature again rose, attaining a point above its previous maximum, this last rise being $0\cdot048°$ C. ($0\cdot0864°$ F.) above the original starting-point.

CHAPTER III.

COMPARATIVE EFFECT OF INTELLECTUAL WORK ON THE TEMPERATURE OF DIFFERENT SPACES OF ONE AND THE SAME SIDE OF THE HEAD.

IT is now time to consider more fully the comparative effect of intellectual work on the temperature of different points of the head. For purposes of comparison the following spaces have been selected in each region :

Anterior region.

5th district, 3rd tier ; 5th district, 1st—2nd tier ; 3rd district, 3rd tier ; 2nd district, 2nd tier ; 3rd—4th district, 5th tier.

Middle region.

1st—2nd district, 4th tier ; 3rd district, 5th tier ; 3rd district, 2nd tier ; 3rd district, 4th tier ; 4th district, 7th tier.

Posterior region.

2nd district, 4th tier ; 5th district, 3rd tier ; 4th district, 2nd tier ; 2nd district, 6th tier.

One of the above spaces,—1st—2nd district, 4th tier, middle region,—is compared, in turn, with each of the other spaces, under circumstances as nearly as possible identical. A pile is placed on each of the two spaces compared, the two currents being opposed to each other, and the deflections of the galvanometer indicating, therefore, the difference of rise of temperature in the two points, admitting that both points have their temperatures raised. The work performed was, as far as possible, the same in the different experiments, and each experiment lasted an hour. The following is a synopsis of the results obtained :

Comparative Effect of Intellectual Work. 147

Comparison of anterior and middle regions.

1st set of experiments.

Mean of 10 observations; 5 on each side of the head.

Comparison of {1st—2nd district, 4th tier, middle region.
5th district, 3rd tier, anterior region.}

Rise of temperature of 0·006° C. (0·0108° F.) in favour of middle region.

2nd set of experiments.

Mean of 10 observations; 4 on right side, 6 on left side.

Comparison of {1st—2nd district, 4th tier, middle region.
5th district, 1st—2nd tier, anterior region.}

Rise of temperature of 0·017° C. (0·0306° F.) in favour of middle region.

3rd set of experiments.

Mean of 10 observations; 5 on each side.

Comparison of {1st—2nd district, 4th tier, middle region.
3rd district, 3rd tier, anterior region.}

Rise of temperature of 0·017° C. (0·0306° F.) in favour of middle region.

4th set of experiments.

Mean of 10 observations; 4 on right side, 6 on left side.

Comparison of {1st—2nd district, 4th tier, middle region.
3rd—4th district, 5th tier, anterior region.}

Rise of temperature of 0·018° C. (0·0324° F.) in favour of middle region.

5th set of experiments.

Mean of 10 observations; 5 on each side.

Comparison of {1st—2nd district, 4th tier, middle region.
2nd district, 2nd tier, anterior region.}

Rise of temperature of 0·02° C. (0·036° F.) in favour of middle region.

Comparison of 1st—2nd district, 4th tier, middle region, with other paces of the same region.

1st set of experiments.

Mean of 10 observations ; 5 on each side.

Comparison of $\begin{cases} \text{1st—2nd district, 4th tier.} \\ \text{3rd district, 4th tier.} \end{cases}$

Rise of temperature of 0·005° C. (0·009° F.) in favour of 1st—2nd district, 4th tier.

2nd set of experiments.

Mean of 10 observations ; 5 on each side.

Comparison of $\begin{cases} \text{1st—2nd district, 4th tier.} \\ \text{3rd district, 2nd tier.} \end{cases}$

Rise of temperature of 0·016° C. (0·0288° F.) in favour of 1st—2nd district, 4th tier.

3rd set of experiments.

Mean of 10 observations ; 4 on right side, 6 on left side.

Comparison of $\begin{cases} \text{1st—2nd district, 4th tier.} \\ \text{3rd district, 5th tier.} \end{cases}$

Rise of temperature of 0·018° C. (0·0324° F.) in favour of 1st—2nd district, 4th tier.

4th set of experiments.

Mean of 10 observations ; 4 on right side, 6 on left side.

Comparison of $\begin{cases} \text{1st—2nd district, 4th tier.} \\ \text{4th district, 7th tier.} \end{cases}$

Rise of temperature of 0·018° C. (0·0324° F.) in favour of 1st—2nd district, 4th tier.

Comparative Effect of Intellectual Work. 149

Comparison of middle and posterior regions.

1st set of experiments.

Mean of 10 observations; 4 on right side, 6 on left side.

Comparison of $\begin{cases} \text{1st—2nd district, 4th tier, middle region.} \\ \text{5th district, 3rd tier, posterior region.} \end{cases}$

Rise of temperature of 0·0195° C. (0·0351° F.) in favour of middle region.

2nd set of experiments.

Mean of 10 observations; 5 on each side.

Comparison of $\begin{cases} \text{1st—2nd district, 4th tier, middle region.} \\ \text{4th district, 2nd tier, posterior region.} \end{cases}$

Rise of temperature of 0·023° C. (0·0414° F.) in favour of middle region.

3rd set of experiments.

Mean of 10 observations; 5 on each side.

Comparison of $\begin{cases} \text{1st—2nd district, 4th tier, middle region.} \\ \text{2nd district, 4th tier, posterior region.} \end{cases}$

Rise of temperature of 0·026° C. (0·0468° F.) in favour of middle region.

4th set of experiments.

Mean of 10 observations; 5 on each side.

Comparison of $\begin{cases} \text{1st—2nd district, 4th tier, middle region.} \\ \text{2nd district, 6th tier, posterior region.} \end{cases}$

Rise of temperature of 0·027° C. (0·0486° F.) in favour of middle region.

It will be seen from the above results, that the 1st—2nd district, 4th tier, middle region, shows a higher rise of temperature than any of the spaces with which it is compared. Taking the value of the absolute rise in this space as 1000, we have the following comparative values for the other spaces :

Values of absolute rises of temperature.

Anterior region.

5th district, 3rd tier	. . . .	846.
5th ,, 1st—2nd tier	. . .	564.
3rd ,, 3rd tier	. . . .	564.
3rd—4th district, 5th tier	. . .	538.
2nd district, 2nd tier	. . . .	513.

Middle region.

1st—2nd district, 4th tier	. . .	1000.
3rd district, 4th tier	. . . .	872.
3rd ,, 2nd ,,	. . . .	589.
3rd ,, 5th ,,	. . . .	538.
4th ,, 7th ,,	. . . .	538.

Posterior region.

5th district, 3rd tier	. . . .	500.
4th ,, 2nd ,,	. . . .	410.
2nd ,, 4th ,,	. . . .	333.
2nd ,, 6th ,,	. . . .	308.

The above order of superiority of rise of temperature in the different spaces is confirmed in the majority of cases by direct mutual comparison of the spaces. The next highest rises after that of the 1st—2nd district, 4th tier, middle region, are found in the 3rd district, 4th tier, of the same region, and in the 5th district, 3rd tier, anterior region. Extended investigations in the tract of highest rise of temperature formed by these spaces and in the neighbouring spaces, have led to the conclusion that the greatest elevations of temperature found at the surface of the head during intellectual work *of all kinds*, are usually met with in the following spaces :

Anterior region.—4th and 5th districts, 3rd and 4th tiers.

Middle region.—1st and 2nd districts, 3rd tier ; 1st, 2nd, and 3rd districts, 4th tier ; 1st and 2nd districts, 5th tier.

The above spaces lie over a tract of the brain-surface formed by the posterior portions of the 1st and 2nd frontal, and the

Comparative Effect of Intellectual Work. 151

anterior ascending parietal [4th frontal] convolutions; and, possibly—crossing the fissure of Rolando—the anterior part of the posterior ascending parietal [ascending parietal] convolution.

We will now proceed to examine, in detail, some of the experiments made by directly comparing two spaces in one and the same observation, on different subjects.

Comparison of anterior and middle regions.

1st Experiment.

Comparison of 1st—2nd district, 4th tier, middle region, with 5th district, 3rd tier, anterior region; both spaces on left side.

+ signifies in favour of middle region.
− signifies in favour of anterior region.

Time from commencement of work.	Rise of temperature.		Mental condition.
	Deflections of galvanometer.	Thermometric values.	
At the end of—			
0 minutes	0°	0°	Commenced mathematical calculations.
2½ ,,	−4°	−0·008° C. (0·0144° F.)	
4½ ,,	−5°	−0·010° C. (0·018° F.)	
5½ ,,	−4°	−0·008° C. (0·0144° F.)	
12½ ,,	+2°	+0·004° C. (0·0072° F.)	
14½ ,,	+3·5°	+0·007° C. (0·0126° F.)	
15½ ,,	+4°	+0·008° C. (0·0144° F.)	
24½ ,,	+5°	+0·010° C. (0·018° F.)	
32½ ,,	+4°	+0·008° C. (0·0144° F.)	
34½ ,,	+ ,,	+ ,, ,,	
39½ ,,	+ ,,	+ ,, ,,	
42½ ,,	+6°	+0·012° C. (0·0216° F.)	
			Stopped work.
47½ ,,	+3°	+0·006° C. (0·0108° F.)	
48 ,,	+2°	+0·004° C. (0·0072° F.)	
			Commenced mathematics again.
49½ ,,	+5°	+0·010° C. (0·018° F.)	
52½ ,,	+6°	+0·012° C. (0·0216° F.)	
53½ ,,	+7°	+0·014° C. (0·0252° F.)	
57½ ,,	+8°	+0·016° C. (0·0288° F.)	

Highest rise, in favour of middle region, 0·016° C. (0·0288° F.).

In this experiment it will be observed, in the first place, that the superiority of rise of temperature of the space of the middle region is much greater than the average given as the result of 10 comparisons between the same spaces in the former experiments. It will also be observed that, during the first five and a half minutes, the higher rise of temperature was in favour of the anterior region. Nor was this last result a negative one, owing to lack of interest or want of attention;—on the contrary, the mental activity was fully as great at this period as at any other of the experiment.

It will be shown, in another place, that two spaces may, under circumstances as nearly as possible identical, exhibit, at different times, a totally different comparative rise of temperature,—the space which, in the one case, exhibited the greater rise, in the other case, exhibiting the lesser rise. In the experiment just given, it would seem probable that such an alternation of activity as that alluded to took place. In the first few minutes the space of the anterior region did the greater share of the work, but after this time the space of the middle region surpassed it. Considering the marked difference of temperature which existed at the conclusion of the experiment, it would further seem not unlikely that, after its first display of activity, the space of the anterior region really performed little or no work at all, the space of the middle region being then to all intents and purposes alone active.

2nd Experiment.

Comparison of 1st—2nd district, 4th tier, middle region, with 3rd—4th district, 5th tier, anterior region; both spaces on left side.

+ signifies in favour of middle region.
− signifies in favour of anterior region.

Time from commencement of work.	Rise of temperature.		Mental condition.
	Deflections of galvanometer.	Thermometric values.	
At the end of—			
0 minutes	0°	0°	Commenced copying from a book, not very attentively.
4 ,,	+1°	+0·002° C. (0·0036° F.)	
7½ ,,	+ ,,	+ ,, ,,	
			Stopped copying.
8½ ,,	−1°	−0·002° C. (0·0036° F.)	
9 ,,	− ,,	− ,, ,,	
10 ,,	− ,,	− ,, ,,	

Comparative Effect of Intellectual Work.

Time from commencement of work.	Rise of temperature.		Mental condition.
	Deflections of galvanometer.	Thermometric values.	
At the end of—			
11 minutes	0°	0°	
12 ,,	0°	0°	
16½ ,,	+3°	+0·006° C. (0·0108° F.)	Commenced copying again attentively.
20 ,,	+8·5°	+0·017° C. (0·0306° F.)	
22 ,,	+7°	+0·014° C. (0·0252° F.)	Stopped copying.
24 ,,	+4°	+0·008° C. (0·0144° F.)	
28½ ,,	+6°	+0·012° C. (0·0216° F.)	Commenced mathematical calculations.
33 ,,	+2°	+0·004° C. (0·0072° F.)	} Interest flagging.
35½ ,,	+3°	+0·006° C. (0·0108° F.)	
38 ,,	+6°	+0·012° C. (0·0216° F.)	Interest returning.
42½ ,,	+7·5°	+0·015° C. (0·027° F.)	
44½ ,,	+9°	+0·018° C. (0·0324° F.)	
48 ,,	+9·5°	+0·019° C. (0·0342° F.)	
50 ,,	+10·5°	+0·021° C. (0·0378° F.)	
			Stopped work.
51 ,,	+7·5°	+0·015° C. (0·027° F.)	
51½ ,,	+7°	+0·014° C. (0·0252° F.)	
52½ ,,	+5·5°	+0·011° C. (0·0198° F.)	

Highest rise, in favour of middle region, 0·021°C. (0·0378°F.).

At the commencement of the above experiment the work was performed carelessly, hence no particular effect was produced. A second attempt was more successful, a rise in favour of the middle region of 0·017° C. (0·0306° F.) being obtained in eight minutes. During four minutes' rest, the excess in favour of the middle region was reduced to 0·008° C. (0·0144° F.). Mathematical work was then begun, causing a rise of 0·013° C. (0·0234° F.) in twenty-six minutes, at the end of which time the maximum rise was attained.

3rd Experiment.

Comparison of 1st—2nd district, 4th tier, middle region, with 5th district, 1st—2nd tier, anterior region; both spaces on left side.

+ signifies in favour of middle region.

Temperature of the Head.

Time from commencement of work.	Rise of temperature.		Mental condition.
	Deflections of galvanometer.	Thermometric values.	
At the end of—			
0 minutes	0°	0°	Commenced copying from book very attentively; much interested. Stopped work.
5½ ,,	+17°	+0·034° C. (0·0612° F.)	
7½ ,,	+14·5°	+0·029° C. (0·0522° F.)	
9 ,,	+12°	+0·024° C. (0·0432° F.)	
12⅔ ,,	+7·5°	+0·015° C. (0·027° F.)	
13 ,,	+7°	+0·014° C. (0·0252 F.)	
			Commenced copying again.
18½ ,,	+17°	+0·034° C. (0·0612° F.)	

Highest rise, in favour of middle region, 0·034°C. (0·0612° F.).

In this experiment five and a half minutes' copying caused a rise in favour of the middle region of 0·034° C. (0·0612° F.). The difference between the temperatures of the two spaces caused by the work was so great, that it would seem not unlikely that the rise was principally, if not almost exclusively, confined to the space of the middle region. The rapid and decided effect produced in this experiment by mental work of a low order seems to have been due to the great interest felt by the subject of the experiment in performing the task in the best possible manner.

Comparison of the 1st—2nd district, 4th tier, middle region, with other spaces of the same region.

1st Experiment.

Comparison of 1st—2nd district, 4th tier, with 3rd district, 5th tier; both spaces on left side.

+ signifies in favour of 1st—2nd district, 4th tier.
− signifies in favour of 3rd district, 5th tier.

At the end of—

0 minutes	0°	0°		Commenced mathematical work.
3 ,,	+1°	+0·002° C. (0·0036° F.)		
7 ,,	+3°	+0·006° C. (0·0108° F.)		
12 ,,	+4·5°	+0·009° C. (0·0162° F.)		
14½ ,,	+ ,,	+ ,,	,,	
16⅔ ,,	+5°	+0·010° C. (0·018° F.)		

Comparative Effect of Intellectual Work.

Time from commencement of work.	Rise of temperature.		Mental condition.
	Deflections of galvanometer.	Thermometric values.	

At the end of—

20 minutes	+4·5°.	+0·009° C. (0·0162° F.)	
25 ,,	+ ,,	+ ,, ,,	
30 ,,	+6°.	+0·012° C. (0·0216° F.)	
35 ,,	+7°.	+0·014° C. (0·0252° F.)	
37½ ,,	+6·5°.	+0·013° C. (0·0234° F.)	
41 ,,	+7°.	+0·014° C. (0·0252° F.)	
44 ,,	+7·5°.	+0·015° C. (0·027° F.)	
50 ,,	+8°.	+0·016° C. (0·0288° F.)	
54 ,,	+7°.	+0·014° C. (0·0252° F.)	
57 ,,	+8·5°.	+0·017° C. (0·0306° F.)	
61 ,,	+9·5°.	+0·019° C. (0·0342° F.)	
65 ,,	+8·5°.	+0·017° C. (0·0306° F.)	Stopped work.
74 ,,	+6°.	+0·012° C. (0·0216° F.)	
81 ,,	+5°.	+0·01° C. (0·018° F.)	
83 ,,	+4°.	+0·008° C. (0·0144° F.)	
85 ,,	+2°.	+0·004° C. (0·0072° F.)	
86 ,,	−1°.	−0·002° C. (0·0036° F.)	

Highest rise, in favour of 1st—2nd district, 4th tier, 0·019° C. (0·0342° F.).

The above experiment requires no special comment. Both spaces were proved to have risen in temperature, by comparing their temperatures before and during the experiment with a constant source of heat.

2nd Experiment.

Comparison of 1st—2nd district, 4th tier, with 3rd district, 4th tier; both spaces on left side.

+ signifies in favour of 1st—2nd district.
− signifies in favour of 3rd district.

At the end of—

0 minutes	0°.	0°	Commenced mathematical calculations; attention strongly concentrated on the work.
3 ,,	+1°.	+0·002° C. (0·0036° F.)	
5 ,,	−2°.	−0·004° C. (0·0072° F.)	
7½ ,,	−3°.	−0·006° C. (0·0108° F.)	
8 ,,	0°.	0°	
10 ,,	+3°.	+0·006° C. (0·0108° F.)	
12½ ,,	+2°.	+0·004° C. (0·0072° F.)	
15 ,,	+ ,,	+ ,, ,,	

156   *Temperature of the Head.*

	Rise of temperature.		
Time from commencement of work.	Deflections of galvanometer.	Thermometric values.	Mental condition.

At the end of—

19 minutes	$+3\cdot5°$	$+0\cdot007°$ C. ($0\cdot0126°$ F.)	
21 ,,	$0°$	$0°$	
$22\frac{1}{2}$,,	$-1°$	$-0\cdot002°$ C. ($0\cdot0036°$ F.)	
26 ,,	$-2\cdot5°$	$-0\cdot005°$ C. ($0\cdot009°$ F.)	
29 ,,	$+2°$	$+0\cdot004°$ C. ($0\cdot0072°$ F.)	
33 ,,	$+4°$	$+0\cdot008°$ C. ($0\cdot0144°$ F.)	
$34\frac{1}{2}$,,	$+$,,	$+$,,	,,
37 ,,	$+3\cdot5°$	$+0\cdot007°$ C. ($0\cdot0126°$ F.)	
			Stopped work.
40 ,,	$+1°$	$+0\cdot002°$ C. ($0\cdot0036°$ F.)	
$44\frac{1}{2}$,,	$0°$	$0°$.	

Highest rise, in favour of 1st—2nd district, $0\cdot008°$ C. ($0\cdot0144°$ F.).

Highest rise, in favour of 3rd district, $0\cdot006°$ C. ($0\cdot0108°$ F.).

In this experiment, the deflections of the galvanometer show a struggle for supremacy of rise of temperature in the two spaces, each one alternately rising above the other, the maximum, however, of all the separate rises of temperature being in favour of the 1st—2nd district. The mental activity was greater than usual during this experiment, and an examination of the temperature of the 3rd district, before and during the experiment, showed that the temperature of this district rose as much as $0\cdot038°$ C. ($0\cdot0684°$ F.); of course the rise in the 1st—2nd district must have been greater still.

3rd Experiment.

Comparison of 1st—2nd district, 4th tier, with 3rd district, 2nd tier; both spaces on left side.

   $+$ signifies in favour of 1st—2nd district, 4th tier.
   $-$ signifies in favour of 3rd district, 2nd tier.

At the end of—

0 minutes	$0°$	$0°$	Commenced reading aloud.
2 ,	$-2°$	$-0\cdot004°$ C. ($0\cdot0072°$ F.)	
5 ,,	$+3°$	$+0\cdot006°$ C. ($0\cdot0108°$ F.)	
8 ,,	$+5°$	$+0\cdot010°$ C. ($0\cdot018°$ F.)	
9 ,,	$+4°$	$+0\cdot008°$ C. ($0\cdot0144°$ F.)	

Comparative Effect of Intellectual Work. 157

Time from commencement of work.	Rise of temperature. Deflections of galvanometer.	Thermometric values.	Mental condition.
At the end of—			
12 minutes	$+3\cdot5°$	$+0\cdot007°$ C. ($0\cdot0126°$ F.)	} Interest flagging.
15 ,,	$+2°$	$+0\cdot004°$ C. ($0\cdot0072°$ F.)	
			Stopped reading.
17 ,,	$+2°$	$+0\cdot004°$ C. ($0\cdot0072°$ F.)	
18 ,,	$+1\cdot5°$	$+0\cdot003°$ C. ($0\cdot0054°$ F.)	
$19\frac{1}{2}$,,	$0°$	$0°$	
			Reading resumed with more interest.
24 ,,	$+4°$	$+0\cdot008°$ C. ($0\cdot0144°$ F.)	
30 ,,	$+6°$	$+0\cdot012°$ C. ($0\cdot0216°$ F.)	
33 ,,	$+7\cdot5°$	$+0\cdot015°$ C. ($0\cdot027°$ F.)	
35 ,,	$+8\cdot5°$	$+0\cdot017°$ C. ($0\cdot0306°$ F.)	
			Stopped reading.
40 ,,	$+7°$	$+0\cdot014°$ C. ($0\cdot0252°$ F.)	
45 ,,	$+5°$	$+0\cdot010°$ C. ($0\cdot018°$ F.)	
47 ,,	$+3°$	$+0\cdot006°$ C. ($0\cdot0108°$ F.)	

Highest rise, in favour of 1st—2nd district, 4th tier, $0\cdot017°$ C. ($0\cdot0306°$ F.).

Both spaces were proved to have had their temperatures raised in the above experiment, the 1st—2nd district, 4th tier, being much more active, however, in this respect than the other space, as the above results show.

Comparison of middle and posterior regions.

1st Experiment.

Comparison of 1st—2nd district, 4th tier, middle region, with 2nd district, 4th tier, posterior region; both spaces on left side.

$+$ signifies in favour of middle region.

At the end of—			
0 minutes	$0°$	$0°$	Commenced mathematical calculations.
$1\frac{1}{2}$,,	$+2°$	$+0\cdot004°$ C. ($0\cdot0072°$ F.)	
3 ,,	$+2\cdot5°$	$+0\cdot005°$ C. ($0\cdot009°$ F.)	
7 ,,	$+4°$	$+0\cdot008°$ C ($0\cdot0144°$ F.)	
9 ,,	$+8°$	$+0\cdot016°$ C. ($0\cdot0288°$ F.)	
13 ,,	$+9\cdot5°$	$+0\cdot019°$ C. ($0\cdot0342°$ F.)	
$16\frac{1}{2}$,,	$+8\cdot5°$	$+0\cdot017°$ C. ($0\cdot0306°$ F.)	
17 ,,	$+9°$	$+0\cdot018°$ C. ($0\cdot0324°$ F.)	

158   *Temperature of the Head.*

Time from commencement of work.	Rise of temperature.		Mental condition.
	Deflections of galvanometer.	Thermometric values.	

At the end of—

19 minutes	. +10°	. +0·02° C. (0·036° F.)	
25 ,,	. +13°	. +0·026° C. (0·0468° F.)	
29 ,,	. +15°	. +0·03° C. (0·054° F.)	
33 ,,	. +17·5°	+0·035° C. (0·063° F.)	
35 ,,	. +14°	. +0·028° C. (0·0504° F.)	
			Stopped work.
38 ,,	. +10°	. +0·02° C. (0·036° F.)	
43 ,,	. +6°	. +0·012° C. (0·0216° F.)	
49 ,,	. +2°	. +0·004° C. (0·0072° F.)	

Highest rise, in favour of middle region, 0·035 C. (0·063° F.).

In this experiment both spaces rose in temperature in a degree decidedly above the average; the rise was much greater, however, in the middle region than in the posterior region, as indicated by the above figures.

2nd Experiment.

Comparison of 1st—2nd district, 4th tier, middle region, with 5th district, 3rd tier, posterior region; both spaces on left side.

+ signifies in favour of middle region.

At the end of—

0 minutes	. 0°	. 0°	Commenced reading aloud.
3 ,,	. +3°	. +0·006° C. (0·0108° F.)	
5 ,,	. +4°	. +0·008° C. (0·0144° F.)	
7½ ,,	. +5°	. +0·01° C. (0·018° F.)	
12 ,,	. +6°	. +0·012° C. (0·0216° F.)	
15 ,,	. + ,,	. + ,,	
21 ,,	. +7°	. +0·014° C. (0·0252° F.)	
26½ ,,	. +7·5°	. +0·015° C. (0·027° F.)	
29 ,,	. +8°	. +0·016° C. (0·0288° F.)	
29⅔ ,,	. +10°	. +0·02° C. (0·036° F.)	
35 ,,	. +9°	. +0·018° C. (0·0324° F.)	
37 ,,	. +9·5°	. +0·019° C. (0·0342° F.)	
			Stopped reading.
45 ,,	. +6°	. +0·012° C. (0·0216° F.)	
47 ,,	. +4°	. +0·008° C. (0·0144° F.)	

Highest rise, in favour of middle region, 0·02° C. (0·036° F.).

In this experiment both spaces had their temperatures raised, the middle region showing the greater rise.

CHAPTER IV.

REVERSALS OF THE USUAL ORDER OF SUPERIORITY OF RISE OF TEMPERATURE IN DIFFERENT SPACES IN INTELLECTUAL WORK.—EFFECT OF VERBAL EXPRESSION IN INTELLECTUAL WORK ON THE COMPARATIVE RISE OF TEMPERATURE IN THE SURFACE OVER BROCA'S CONVOLUTION.— PRECAUTIONS RESPECTING THE ABSOLUTE TEMPERATURES OF THE PARTS EXAMINED.

It has been stated (p. 152) that two spaces compared at different times may show a reversal of the order of supremacy of rise of temperature, the space which, in one instance, exhibited the greater rise, in the other instance, showing the lesser rise. Such examples are most frequent among the spaces which have been cited as showing the greatest elevations of temperature. An example of superiority of rise of temperature alternating between two spaces, in one and the same observation, is given in the comparison of 1st—2nd district, 4th tier, and 3rd district, 4th tier, middle region (p. 155). The following experiments belong to the same class as this latter.

1st Experiment.

Comparison of 1st—2nd district, 4th tier, middle region, with 5th district, 3rd tier, anterior region; both spaces on left side.

$+$ signifies in favour of middle region.
$-$ signifies in favour of anterior region.

Time from commencement of work.	Rise of temperature.		Mental condition.
	Deflections of galvanometer.	Thermometric values.	
At the end of—			
0 minutes	$0°$	$0°$	Commenced mathematical calculations.
11 ,,	$-3°$	$-0·006°$ C. ($0·0108°$ F.)	
$17\frac{1}{2}$,,	$-2°$	$-0·004°$ C. ($0·0072°$ F.)	
20 ,,	$0°$	$0°$	
40 ,,	$+3°$	$+0·006°$ C. ($0·0108°$ F.)	

160   *Temperature of the Head.*

Time from
commence- Deflections of Thermometric
ment of work. galvanometer.   values.   Mental condition.

Rise of temperature.

At the end of—
50 minutes . +5°   . +0·01° C. (0·018° F.)
55   ,,   . +5·5° . +0·011° C. (0·0198° F.)
60   ,,   . +2·5° . +0·005° C. (0·009° F.)
70   ,,   . +7°   . +0·014° C. (0·0252° F.)
75   ,,   . +4°   . +0·008° C. (0·0144° F.)
76½   ,,   . +1°   . +0·002° C. (0·0036° F.)
78½   ,,   . −1°   . −0·002° C. (0·0036° F.)
79½   ,,   . −2°   . −0·004° C. (0·0072° F.)

Highest rise, in favour of middle region, 0·014° C. (0·0252° F.).
    ,,   ,,   anterior ,,   0·006° C. (0·0108° F.).

Here the rise of temperature is alternately greater in each space, the highest total rise being, however, still—in accordance with the rule—in favour of the middle region.

2nd Experiment.

Comparison of 1st—2nd district, 4th tier, middle region, with 5th district, 3rd tier, anterior region; both spaces on left side.

+ signifies in favour of middle region.
− signifies in favour of anterior region.

At the end of—
0 minutes . 0°   .   0°   Commenced mathematical
5½   ,,   . ,,   . ,,   calculations.
12   ,,   . −1°   . −0·002° C. (0·0036° F.)
16   ,,   . −4°   . −0·008° C. (0·0144° F.)
17½ ,,   . −1°   . −0·002° C. (0·0036° F.)
22   ,,   . −3°   . −0·006° C. (0·0108° F.)
25   ,,   . −5°   . −0·01° C. (0·018° F.)
30½ ,,   . −3°   . −0·006° C. (0·0108° F.)
35   ,,   . −5·5° . −0·011° C. (0·0198° F.)
40   ,,   . −,, . −,,   ,,
42½ ,,   . −6·5° . −0·013° C. (0·0234° F.)
48   ,,   . −6°   . −0·012° C. (0·0216° F.)
48½ ,,   . −5°   . −0·01° C. (0·018° F.)
                                            Stopped work.
49½ ,,   . −4°   . −0·008° C. (0·0144° F.)

At this point both the work and the observation of the temperature ceased for about half an hour. At the end of this time a new zero was taken, and the observations again commenced.

Comparative Effect of Intellectual Work.

Time from commencement of work.	Deflections of galvanometer.	Rise of temperature. Thermometric values.	Mental condition.
At the end of—			
0 minutes	0°	0°	Commenced mathematical calculations.
1 ,,	+1°	+0·002° C. (0·0036° F.)	
3 ,,	+2°	+0·004° C. (0·0072° F.)	
11½ ,,	+3°	+0·006° C. (0·0108° F.)	
15 ,,	+ ,,	+ ,,	,,
22½ ,,	+5°	+0·01° C. (0·018° F.)	
26½ ,,	+ ,,	+ ,,	,,
31 ,,	+6°	+0·012° C. (0·0216° F.)	
39 ,,	+7°	+0·014° C. (0·0252° F.)	
42 ,,	+6·5°	+0·013° C. (0·0234° F.)	
			Stopped work.
44 ,,	+4°	+0·008° C. (0·0144° F.)	
45 ,,	+2°	+0·004° C. (0·0072° F.)	
46 ,,	+1°	+0·002° C. (0·0036° F.)	

In the first part of the above experiment the rise of temperature was greater in the anterior region than in the middle region. After half an hour's interval of rest the same kind of work as that employed in the first part of the experiment produced a greater rise in the middle region than in the anterior region. The superiority of rise of one space over the other was nearly the same in the two cases; the rise in favour of the anterior region being 0·013° C. (0·0234° F.), while that in favour of the middle region was 0·014° C. (0·0252° F.). The work was about equally difficult in the two cases, and the mind appeared to have about the same degree of activity in both instances.

3rd Experiment.

Comparison of 5th district, 3rd tier, anterior region, and 4th district, 4th tier, same region; both spaces on left side.

+ signifies in favour of 4th district, 4th tier.

At the end of—			
0 minutes	0°	0°	Commenced mathematical calculations.
2 ,,	+4°	+0·008° C. (0·0144° F.)	
6 ,,	+7°	+0·014° C. (0·0252° F.)	
10 ,,	+18°	+0·036° C. (0·0648° F.)	
11½ ,,	+20°	+0·04° C. (0·072° F.)	
			Stopped work.
13 ,,	+16°	+0·032° C. (0·0576° F.)	

Temperature of the Head.

Time from commencement of work.	Rise of temperature.		Mental condition.
	Deflections of galvanometer.	Thermometric values.	

At the end of—

14 minutes	+14°	+0·028° C. (0·0504° F.)	
14½ ,,	+13°	+0·026° C. (0·0468° F.)	
18 ,,	+13°	+0·026° C. (0·0468° F.)	
21 ,,	+12°	+0·024° C. (0·0432° F.)	
27 ,,	+24°	+0·048° C. (0·0864° F.)	
30½ ,,	+31°	+0·062° C. (0·1116° F.)	
			Stopped work.
34 ,,	+24°	+0·048° C. (0·0864° F.)	

Highest rise, in favour of 4th district, 4th tier, 0·062° C. (0·1116° F.).

The unusually great difference in rise of temperature between the two spaces in this case points to local vascular disturbance in the 4th district, 4th tier. It was found that the 5th district, 3rd tier, had risen as well as the other space, although the exact degree of rise was not known. The extreme rise in the 4th district, 4th tier, seems to be explainable only in the way mentioned.

In forty-eight comparisons of the 1st—2nd district, 4th tier, middle region, and the 5th district, 3rd tier, anterior region, in which the experimental conditions were as nearly as possible the same, thirty-three results were in favour of the space of the middle region, twelve results were in favour of the space of the anterior region, and in the remaining three cases the rise was equal in the two spaces.

But the spaces in which the highest rises of temperature are usually found may sometimes be surpassed in thermal activity by spaces not belonging to their set. It is not unfrequently the case that the 5th district, 1st—2nd tier, anterior region, shows a higher rise of temperature than the 1st—2nd district, 4th tier, middle region. The space specified of the anterior region is over the posterior portion of the 3rd frontal convolution.

An attempt has been made to see if, in the comparison of the above-named two spaces, the use of articulate language in the work performed has any effect on the position of the higher rise of temperature.

When the possibility of the existence of a connection between

Comparative Effect of Intellectual Work.   163

the superiority of rise of temperature in the space in question of the anterior region, and the marked use of articulate language, first suggested itself, two classes of experiments were being made, at the same time, on the two spaces under consideration, on two persons. With one individual, the work consisted in the very careful reading aloud of matter which required considerable thought. The reader both understood, and was much interested in the matter read, and also made a strong effort to read effectively. With the other individual, the work was the same except that there was no verbal expression.

In comparing the results obtained on these two persons, it was noticed that the higher rise of temperature was more frequently found in the space of the anterior region, in the case of the individual reading aloud, than in the case of the other person. On reversing the conditions of the experiment—each individual having now the work of the other—there appeared, at first, to be reason to believe that the position of higher rise of temperature in the two persons had been correspondingly affected.

The following are examples of the experiments made specially to see if reading aloud had any particular effect upon the position of higher rise of temperature in the comparison of the spaces in question :—

1st Experiment.

$+$ signifies in favour of middle region.
$-$ signifies in favour of anterior region.

Time from commencement of work.	Rise of temperature.		Mental condition.
	Deflections of galvanometer.	Thermometric values.	
At the end of—			
0 minutes	$0°$	$0°$	Commenced reading aloud with care; also interested in matter read.
$4\frac{1}{2}$,,	$-1°$	$-0·002°$ C. ($0·0036°$ F.)	
$19\frac{1}{2}$,,	$-2°$	$-0·004°$ C. ($0·0072°$ F.)	
20 ,,	$-1·5°$	$-0·003°$ C. ($0·0054°$ F.)	
			Stopped reading.
22 ,,	$-1°$	$-0·002°$ C. ($0·0036°$ F.)	
24 ,,	$-$,,	$-$,, ,,	
25 ,,	$0°$	$0°$	
			Commenced reading to one's self.
30 ,,	$+2°$	$+0·004°$ C. ($0·0072°$ F.)	
35 ,,	$+3°$	$+0·006°$ C. ($0·0108°$ F.)	

Time from commencement of work.	Rise of temperature.		Mental condition.
	Deflections of galvanometer.	Thermometric values.	

At the end of—

41½ minutes	+4°	+0·008° C. (0·0144° F.)	
51½ ,,	+5°	+0·01° C. (0·018° F.)	
63½ ,,	+4°	+0·008° C. (0·0144° F.)	
67½ ,,	+ ,,	+ ,, ,,	
68½ ,,	+ ,,	+ ,, ,,	Stopped reading.
70 ,,	+3°	+0·006° C. (0·0108° F.)	
71 ,,	+2°	+0·004° C. (0·0072° F.)	
73 ,,	+1°	+0·002° C. (0·0036° F.)	
74 ,,	0°	0°	

In this experiment the greater rise of temperature was in the anterior region during the reading aloud, and in the middle region during the reading to one's self.

2nd Experiment.

+ signifies in favour of middle region.
− signifies in favour of anterior region.

At the end of—

0 minutes	0°	0°	Commenced mathematical calculations; not interested.
4½ ,,	−8°	−0·016° C. (0·0288° F.)	
17½ ,,	−6°	−0·012° C. (0·0216° F.)	
28 ,,	−1°	−0·002° C. (0·0036° F.)	Interest becoming aroused.
37½ ,,	+4°	+0·008° C. (0·0144° F.)	
40½ ,,	+9°	+0·018° C. (0·0324° F.)	
57½ ,,	+10°	+0·02° C. (0·036° F.)	
65 ,,	+ ,,	+ ,, ,,	Stopped work.
70½ ,,	+9°	+0·018° C. (0·0324° F.)	
73½ ,,	+8°	+0·016° C. (0·0288° F.)	
76½ ,,	+ ,,	+ ,, ,,	Commenced reading aloud very carefully.
90½ ,,	+1°	+0·002° C. (0·0036° F.)	
94½ ,,	−4°	−0·008° C. (0·0144° F.)	Stopped reading.
99½ ,,	+2°	+0·004° C. (0·0072° F.)	
101½ ,,	+4°	+0·008° C. (0·0144° F.)	
106 ,,	+5·5°	+0·011° C. (0·0198° F.)	
110½ ,,	+7·5°	+0·015° C. (0·027° F.)	

In this experiment the movement of the galvanometer needle

in favour of the anterior region, at the commencement of the work, did not (as was proved by direct test) denote a rise of temperature in this region, but was in reality the result of a *fall* of temperature in the middle region; the interest felt in the mathematical work at this time being not only insufficient to raise the temperature, but the degree of mental activity being actually less than that existing before the experiment. Later on, the interest being aroused, a rise of 0·02° C. (0·036° F.) above the zero point, in favour of the middle region, was obtained. After this, reading aloud brought about, in eighteen minutes, a positive rise of 0·008° C. (0·0144° F.) in favour of the anterior region. Seeing, however, that on the cessation of the reading the deflection of the galvanometer passed over to the middle-region side of zero, very nearly attaining its former degree (+8° galv.) when the reading was commenced, it would seem that the reading aloud really caused a rise in the anterior region of 12° galv.=0·024° C. (0·0432° F.) (from +8° to −4° galv.) over that of the middle region. The rise in this last region—if, indeed, any appreciable rise at all occurred—was very slight.

3rd Experiment.

+ signifies in favour of middle region.
− signifies in favour of anterior region.

Time from commencement of work.	Rise of temperature.		Mental condition.
	Deflections of galvanometer.	Thermometric values.	
At the end of—			
0 minutes	0°	0°	Commenced reading aloud:
5 ,,	−3°	−0·006° C. (0·0108° F.)	reading carefully, but not interested in matter read.
8 ,,	−6°	−0·012° C. (0·0216° F.)	
11 ,,	+4°	+0·008° C. (0·0144° F.)	Interested in matter read.
19 ,,	+ ,,	+ ,,	,,
25½ ,,	+7°	+0·014° C. (0·0252° F.)	
32 ,,	+10°	+0·02° C. (0·036° F.)	
39 ,,	+3°	+0·006° C. (0·0108° F.)	Interest flagging. Stopped reading.
44½ ,,	−4°	−0·008° C. (0·0144° F.)	
46 ,,	−8°	−0·016° C. (0·0288° F.)	
47 ,,	−7°	−0·014° C. (0·0252° F.)	

In this experiment the deflection in favour of the anterior

region, at the commencement of the reading, was due to a fall of temperature in the middle region. When the interest was aroused a rise in favour of the middle region manifested itself. There was also a much smaller rise in the anterior region. After the reading ceased, the temperature of the middle region fell below the original starting-point.

4th Experiment.

+ signifies in favour of middle region.
− signifies in favour of anterior region.

Time from commencement of work.	Deflections of galvanometer.	Rise of temperature. Thermometric values.	Mental condition.
At the end of—			
0 minutes	0°	0°	Commenced reading aloud very carefully; decidedly interested in the matter read.
5 ,,	+3°	+0·006° C. (0·0108° F.)	
7 ,,	+5°	+0·01° C. (0·018° F.)	
9½ ,,	+6°	+0·012° C. (0·0216° F.)	
15 ,,	+7°	+0·014° C. (0·0252° F.)	
19 ,,	+6·5°	+0·013° C. (0·0234° F.)	
25 ,,	+7°	+0·014° C. (0·0252° F.)	Stopped reading.
30 ,,	+5·5°	+0·011° C. (0·0198° F.)	
35 ,,	+2·5°	+0·005° C. (0·009° F.)	
37 ,,	+1°	+0·002° C. (0·0036° F.)	Commenced mathematical calculations.
44 ,,	+4°	+0·008° C. (0·0144° F.)	
48 ,,	+6°	+0·012° C. (0·0216° F.)	
55 ,,	+7·5°	+0·015° C. (0·027° F.)	
60 ,,	+ ,,	+ ,, ,,	Stopped work.
65 ,,	+5°	+0·01° C. (0·018° F.)	
69 ,,	+3°	+0·006° C. (0·0108° F.)	
73 ,,	+1·5°	+0·003° C. (0·0054° F.)	Commenced again to read aloud. Reading carefully, and much interested.
77 ,,	+4·5°	+0·009° C. (0·0162° F.)	
82 ,,	+5·5°	+0·011° C. (0·0198° F.)	
88 ,,	+6·5°	+0·013° C. (0·0234° F.)	
90 ,,	+8°	+0·016° C. (0·0288° F.)	Stopped reading.
95 ,,	+5°	+0·01° C. (0·018° F.)	
102 ,,	+3°	+0·006° C. (0·0108° F.)	

In the above experiment, as in the one immediately preceding it, reading aloud caused a higher rise of temperature in the middle region than in the anterior region. These last two results are, therefore, in opposition to the first two.

It must be borne in mind, with reference to these experiments, that in all cases a certain amount of intellectual exertion is necessary—mere mechanical articulation being without effect, as already stated.* This, of course, complicates the problem, as we cannot try the comparative effect on different spaces of simple mechanical reading and speaking,—and the moment intellectual effort is invoked, certain spaces have an advantage over others— so far as averages are concerned—whatever the nature of the work may be.

The alternation, in one and the same experiment, of reading aloud and of reading to one's self, as practised in 1st experiment —keeping the interest, at the same time, at an even pitch— although difficult of execution, seems, nevertheless, upon the whole, to be the most satisfactory method of investigation.

In eighteen cases in which this method has been carried out successfully, thirteen of the results have indicated the greater rise of temperature, both in reading aloud and in reading to one's self, to be in favour of the space of the middle region : in the remaining five cases, the space of the middle region has shown the higher rise during silent reading, and the space of the anterior region the higher rise during reading aloud.

In judging these results it must, however, be borne in mind, that—as we have seen †—a space (5th district, 3rd tier, anterior region) having ordinarily greater thermal activity than the one we are examining, nevertheless furnishes a lower average than the latter has just shown, when compared, like it, with the 1st— 2nd district, 4th tier, middle region. The average for the 5th district, 1st—2nd tier, anterior region, given in the above series of experiments, is greater than that ordinarily obtained in comparison with the space in question of the middle region ; still, this average is hardly enough to justify a conclusion that the temperature of the surface over Broca's convolution is specially affected by reading aloud.

Before leaving this part of our subject it will be well to call attention to a point which concerns all experiments on the comparative rise of temperature in the head. It is—*that the absolute*

* P. 123. † P. 162.

temperatures of the parts experimented on, at the commencement of the observation, be duly regarded. It frequently happens that one region or part is considerably nearer its maximum temperature than the others, at the start. This is especially liable to be the case with the anterior region. Now the nearer the temperature is to the upper limit of thermal range, beyond which no rise can normally take place, the less likelihood is there of the occurrence of an increase of heat in a noticeable degree. Having, therefore, for example, one pile in the anterior region, and the other pile in the posterior region, if the former locality be much the nearer of the two to its maximum temperature, the rise of temperature per unit of time in the posterior region may surpass that in the anterior region, and the deflections of the galvanometer thus lead to wrong conclusions as to the relative amounts of work performed by the two parts. It was, in part, a disregard of this point which led the author in his earlier experiments to consider that the rise of temperature was greater in the posterior region than in the anterior region during mental work.

CHAPTER V.

COMPARATIVE EFFECT OF INTELLECTUAL WORK ON THE TEMPERATURE OF SYMMETRICALLY SITUATED SPACES OF THE TWO SIDES OF THE HEAD.

WE pass now to the comparative effect of intellectual work on the temperature of the two sides of the head.

The spaces selected in each region for examination are those already chosen in the preceding observations,* with the exceptions that, in the middle region, the 4th district, 6th tier, is substituted for the 4th district, 7th tier (the latter space being, on one of its borders, on the median line), and the 3rd district, 4th tier, is omitted.

The comparisons given below were made with two piles with opposed currents, placed one on each side of the head. The

* P. 146.

Comparative Effect of Intellectual Work.   169

work performed was usually one of the four kinds already specified.*

ANTERIOR REGION.

1st Series of Experiments.
Comparison of 3rd district, 3rd tier, on the two sides.
8 observations.

Times of occurrence of superiority of rise of temperature on a side, and of equality of rise of temperature on the two sides.

Mean difference of rise of temperature observed at the end of 1 hour's work.

In favour of—
Left side . 7
Right ,,   . 1

In favour of—
Left side . 0·0115° C. (0·0207° F.)
Right ,, . 0·002° C. (0·0036° F.)

2nd Series of Experiments.
Comparison of 2nd district, 2nd tier, on the two sides.
8 observations.

In favour of—
Left side . 6
Right ,,   . 2

In favour of—
Left side . 0·0045° C. (0·0081° F.)
Right ,, . 0·0018° C. (0·00324° F.)

3rd Series of Experiments.
Comparison of 3rd—4th district, 5th tier, on the two sides.
8 observations.

In favour of—
Left side   . 6
Right ,,   . 1
Equality   . 1

In favour of—
Left side . 0·00525° C. (0·00945° F.)
Right ,, . 0·004° C. (0·0072° F.)

4th Series of Experiments.
Comparison of 5th district, 3rd tier, on the two sides.
8 observations.

In favour of—
Left side   . 5
Right ,,   . 1
Equality   . 2

In favour of—
Left side . 0·003° C. (0·0054° F.)
Right ,, . 0·004° C. (0·0072° F.)

* Pp. 125-127.

Temperature of the Head.

5th Series of Experiments.
Comparison of 5th district, 1st—2nd tier, on the two sides.
8 observations.

Times of occurrence of superiority of rise of temperature on a side, and of equality of rise of temperature on the two sides.

Mean difference of rise of temperature observed at the end of 1 hour's work.

In favour of—
Left side . 5
Right ,, . 2
Equality . 1

In favour of—
Left side . 0·00525° C. (0·00945° F.)
Right ,, . 0·004° C. (0·0072° F.)

MIDDLE REGION.

1st Series of Experiments.
Comparison of 1st—2nd district, 4th tier, on the two sides.
8 observations.

In favour of—
Left side . 6
Right ,, . 1
Equality . 1

In favour of—
Left side . 0·0028° C. (0·00504° F.)
Right ,, . 0·0015° C. (0·0027° F.)

2nd Series of Experiments.
Comparison of 3rd district, 5th tier, on the two sides.
8 observations.

In favour of—
Left side . 6
Right ,, . 1
Equality . 1

In favour of—
Left side . 0·01125° C. (0·02025° F.)
Right ,, . 0·003° C. (0·0054° F.)

3rd Series of Experiments.
Comparison of 3rd district, 2nd tier, on the two sides.
8 observations.

In favour of—
Left side . 5
Right ,, . 2
Equality . 1

In favour of—
Left side . 0·003° C. (0·0054° F.)
Right ,, . 0·003° C. (0·0054° F.)

Comparative Effect of Intellectual Work. 171

4th Series of Experiments.
Comparison of 4th district, 6th tier, on the two sides.
8 observations.

Times of occurrence of superiority of rise of temperature on a side, and of equality of rise of temperature on the two sides.

Mean difference of rise of temperature observed at the end of 1 hour's work.

In favour of—
Left side . 4
Right ,, . 2
Equality . 2

In favour of—
Left side . 0·0028° C. (0·00504° F.)
Right ,, . 0·0015° C. (0·0027° F.)

POSTERIOR REGION.

1st Series of Experiments.
Comparison of 2nd district, 4th tier, on the two sides.
8 observations.

In favour of—
Left side . 4
Right ,, . 2
Equality . 2

In favour of—
Left side . 0·00275° C. (0·00495° F.)
Right ,, . 0·0015° C. (0·0027° F.)

2nd Series of Experiments.
Comparison of 4th district, 2nd tier, on the two sides.
8 observations.

In favour of—
Left side . 5
Right ,, . 1
Equality . 2

In favour of—
Left side . 0·002° C. (0·0036° F.)
Right ,, . 0·0015° C. (0·0027° F.)

3rd Series of Experiments.
Comparison of 2nd district, 6th tier, on the two sides.
8 observations.

In favour of—
Left side . 5
Right ,, . 2
Equality . 1

In favour of—
Left side . 0·0025° C. (0·0045° F.)
Right ,, . 0·0015° C. (0·0027° F.)

4th Series of Experiments.
Comparison of 5th district, 3rd tier, on the two sides.
8 observations.

Times of occurrence of superiority of rise of temperature on a side, and of equality of rise of temperature on the two sides.

In favour of—
Left side . 5
Right ,, . 2
Equality . 1

Mean difference of rise of temperature observed at the end of 1 hour's work.

In favour of—
Left side . 0·002° C. (0·0036° F.)
Right ,, . 0·002° C. (0·0036° F.)

The preceding experiments indicate that, in the greater number of cases, in all three regions, the greater rise of temperature, in intellectual work, takes place on the left side. In a certain number of cases, however, in each space of each region, the rise is in favour of the right side,—and in other instances, in certain spaces, the rise is equal on the two sides. The anterior region shows both the greatest majority of cases in favour of the left side, and also the greatest degree of difference of rise. Next comes the middle region, and last, the posterior region.

The average percentages of times of occurrence of superiority of rise of temperature on a side, and of equality of rise, and the average degree of difference of rise, for all the series of experiments in each region, are as follows:

ANTERIOR REGION.

	Left side.	Right side.	Equality.
Percentages of times of occurrence of superiority of rise of temperature on a side, and of equality of rise of temperature on the two sides	72·500	17·500	10·000
Average degree of difference of rise of temperature	0·0059° C. (0·0106° F.)	0·00316° C. (0·00568° F.)	

MIDDLE REGION.

	Left side.	Right side.	Equality.
Percentages of times of occurrence of superiority of rise of temperature on a side, and of equality of rise of temperature on the two sides . . .	65·625	18·75	15·625

Average degree of difference of rise of temperature } 0·00496° C. (0·00892° C.) . 0·00225° C. (0·0045° F.)

POSTERIOR REGION.

	Left side.	Right side.	Equality.
Percentages of times of occurrence of superiority of rise of temperature on a side, and of equality of rise of temperature on the two sides . . .	59·375	21·875	18·75

Average degree of difference of rise of temperature } 0·00231° C. (0·00415° F.) 0·00162° C. (0·00291° F.)

Of the total 104 observations, 69, or 66·346 per cent., are in favour of the left side; 20, or 19·231 per cent., are in favour of the right side; and 15, or 14·423 per cent., are in favour of equality of the two sides.

When the mental work has been extreme, or very prolonged, and especially if it has been of an exciting character, a class of phenomena sometimes appear which seem to be more or less complicated with vascular disturbance. These phenomena consist principally in irregular shiftings of the balance of superiority of temperature from one side to the other. Under ordinary circumstances, the *position* of the balance of superiority of temperature remains about the same during mental exertion, the *degree* of difference of temperature being alone affected,—that is, if, of two points on different sides, the one "a" have a temperature 0·2° C. higher than the other "b," we will find, after increased mental exertion, "a" still the higher in temperature, but with an increased or diminished difference of degree, according as "a" or "b" has risen in the greater ratio. If the initial difference of temperature be very small, it may be surpassed by the greater ratio of rise of the cooler point, and the position of superiority of temperature thus shifted over to the opposite side. For example, suppose "a" to be warmer than "b" by only $0.008°$ C. ($0.0144°$ F.), and "b" to rise $0.034°$ C. ($0.0612°$ F.), while "a" rises but $0.01°$ C. ($0.018°$ F.);—"b" will now surpass "a" in temperature by $0.016°$ C. ($0.0288°$ F.); the side which was the cooler at the commencement of the experiment is now, therefore, the warmer. Of course everything depends upon the initial difference of temperature, and commonly this is too great for the feeble rise of temperature occurring during mental work to surpass,—but, in the particular cases to which we have alluded, the rise of temperature is much greater, and we see the ordinary order of distribution of temperature first confused, and then frequently almost reversed. These changes are most common in the anterior region, and generally result in an increase in the extent of the tract of *left* superiority of temperature, as well as in an increase in the degree of difference of temperature between the two sides, the balance being in favour of the left side. The middle and posterior regions may be likewise affected, and in the same manner and direction, but the change is, with them, less common, and less marked when it does occur. In the anterior region it is usually the 4th and 5th districts of the 1st, 2nd and 3rd tiers, that have their balances of superiority of temperature shifted over from the right side to the left. In the middle and posterior regions there appears to be no definite order in the alterations of the distribution of superiority of temperature.

CHAPTER VI.

GENERAL EFFECT OF EMOTIONAL ACTIVITY ON THE TEMPERATURE OF THE HEAD.—EXPERIMENTS ILLUSTRATIVE OF THE EFFECT OF EMOTIONAL ACTIVITY.—AVERAGE RISES OF TEMPERATURE FOR THE DIFFERENT REGIONS, IN EMOTIONAL ACTIVITY.

Emotional Activity.

WE have now to consider the last, and, in many respects, the most difficult part of our subject,—namely, the effect of emotional states of the mind on the temperature of the different portions of the surface of the head.

Of these emotional conditions only one class has been found available for strict experimental examination. It is that condition of the mind which is brought about by the reading or recitation aloud, or to one's self, of poetical or prose productions of an emotional character. Many individuals, especially those belonging to the dramatic profession or having a strong taste in that direction, can excite this condition almost at will. Moreover, the mere listening to the effective reading of literature of the kind in question can cause a decided rise of temperature in the head in some persons. In one of the subjects of the present experiments, listening to the reading of poetry was the means usually employed with success to bring about a rise of temperature.

When real emotion is thus aroused, its influence on the temperature of the head shows itself more quickly and is more marked than the influence of intellectual exertion. Mere mechanical reading or recitation, as before stated (p. 123) produces no effect.

In the first class of experiments now to be given, each region was tested separately, as in the corresponding class of experiments on intellectual work, the absolute values of the changes of temperature being thus obtained. So far as it could be done, the work was equalised in the different observations, and the ensuing experiments are selected ones, which were as closely comparable as possible with reference to the circumstances under which they were performed. The spaces selected for examination were the same as those chosen for the study of intellectual work.

ANTERIOR REGION.

Mean of 15 observations on 3rd district, 3rd tier—9 observations on left side, 6 observations on right side. Mean temperature of air 13° C. (55·4° F.). Mean temperature of space examined, at commencement of observations, 33·5° C. (92·3° F.). Reading poetry aloud.

Rise of temperature.

Time from the commencement of reading.	Deflections of galvanometer.	Thermometric values.
At the end of 2½ minutes	8°	0·016° C. (0·0288° F.).
,, 5 ,,	10·4°	0·0208° C. (0·0374° F.).
,, 10 ,,	12·5°	0·025° C. (0·045° F.).
,, 20 ,,	17°	0·034° C. (0·0612° F.).
,, 30 ,,	20°	0·04° C. (0·072° F.).

MIDDLE REGION.

Mean of 15 observations on 3rd district, 5th tier—10 observations on left side, 5 observations on right side. Mean temperature of air 13·2° C. (55·76° F.). Mean temperature of space examined, at commencement of observations, 33·2° C. (91·76° F.). Reading poetry aloud.

At the end of 2½ minutes	7°	0·014° C. (0·0252° F.).
,, 5 ,,	9·5°	0·019° C. (0·0342° F.).
,, 10 ,,	11·6°	0·0232° C. (0·0417° F.).
,, 20 ,,	16·5°	0·033° C. (0·0594° F.).
,, 30 ,,	19·6°	0·0392° C. (0·0705° F.).

POSTERIOR REGION.

Mean of 15 observations on 2nd district, 4th tier—9 observations on left side, 6 observations on right side. Mean temperature of air 13·4° C. (56·12° F.). Mean temperature of space examined,

General Effect of Emotional Activity.

at commencement of observations, 33° C. (91·4° F.). Reading poetry aloud.

Time from the commencement of reading.	Deflections of galvanometer.	Rise of temperature. Thermometric values.
At the end of 2½ minutes	6°	0·012° C. (0·0216° F.)
,, 5 ,,	8·2°	0·0164° C. (0·0295° F.)
,, 10 ,,	10·7°	0·0214° C. (0·0385° F.)
,, 20 ,,	15·2°	0·0304° C. (0·0547° F.)
,, 30 ,,	17·5°	0·035° C. (0·063° F.)

Rises of temperature in preceding experiments.

	Highest rises.	Average rises per minute in thirty minutes.
Anterior region	0·04° C. (0·072° F.)	0·001332° C. (0·002397° F.)
Middle ,,	0·0392° C. (0·0705° F.)	0·001306° C. (0·00235° F.)
Posterior ,,	0·035° C. (0·063° F.)	0·001166° C. (0·002098° F.)

Average highest rise in all three regions taken together.	Average rise per minute in thirty minutes in all three regions taken together.
0·03806° C. (0·0685° F.)	0·001268° C. (0·002282° F.)

The above experiments indicate:—

1st. That emotional activity of the kind in question, like intellectual work, causes a rise of temperature in all three regions of the head; and that this rise is both more rapid and of greater degree than that usually seen in intellectual work.

2nd. That less difference exists in the rapidity and degree of rise of temperature in different regions in emotional activity than in intellectual work; but that the order of the regions, with regard to the comparative degree of rise of temperature, is,—so far as the three spaces in question are concerned,—the same in both cases.

As was found in the experiments on intellectual work, emotional activity may produce a rise of temperature in all parts of the head, but the relative degree of this rise in different spaces exhibits, as a rule, less difference than that usually seen in intellectual exertion. Still, under favorable circumstances, it can be shown that, during emotional activity of the kind specified, the position of the highest rises of temperature is the same as that held in intellectual work. Before, however, comparing in detail the different spaces with each other, examples of the ordinary rises of temperature, due to either reading aloud of poetry by the subject of the experiment, or—as in the case of one individual— to the listening of the subject to the reading of poetry by another person, will be given, as was done at a similar stage, when examining the effects of intellectual work. Two examples from each region are given. As in the corresponding experiments on intellectual work, the " plus " sign indicates a rise above, and the " minus " sign a fall below the starting temperature of the space examined.

ANTERIOR REGION.

1st Experiment.

Examination of 3rd district, 3rd tier, left side.

Time from commence- ment of work.	Rise or fall of temperature.		Mental condition.
	Deflections of galvanometer.	Thermometric values.	
At the end of—			
0 minutes	0°	0°	Reading of poetry to subject commenced.
2 ,,	−1°	−0·002° C. (0·0036° F.)	Not attentive.
4 ,,	+8°	+0·016° C. (0·0288° F.)	Thoroughly interested.
10 ,,	+12°	+0·024° C. (0·0432° F.)	
15 ,,	+15°	+0·030° C. (0·054° F.)	
17 ,,	+16°	+0·032° C. (0·0576° F.)	
18 ,,	+15°	+0·030° C. (0·054° F.)	
22 ,,	+17°	+0·034° C. (0·0612° F.)	
24 ,,	+16°	+0·032° C. (0·0576° F.)	
28 ,,	+17°	+0·034° C. (0·0612° F.)	
33 ,,	+18°	+0·036° C. (0·0648° F.)	
35 ,,	+17·5°	+0·035° C. (0·063° F.)	
38 ,,	+18°	+0·036° C. (0·0648° F.)	

In this experiment the listening to the reading caused a rise of temperature in the head of 0·036° C. (0·0648° F.), in thirty-three minutes.

General Effect of Emotional Activity.

2nd Experiment.
Examination of 5th district, 3rd tier, left side.

Time from commencement of work.	Deflections of galvanometer.	Thermometric values.	Mental condition.
	Rise or fall of temperature.		

At the end of—

0 minutes	0°	0°	Commenced reading aloud poetry.
1 ,,	+3°	+0·006° C. (0·0108° F.)	
2½ ,,	+4°	+0·008° C. (0·0144° F.)	
4 ,,	+3°	+0·006° C. (0·0108° F.)	Stopped reading.
5 ,,	+2°	+0·004° C. (0·0072° F.)	
7 ,,	+1°	+0·002° C. (0·0036° F.)	Commenced reading again.
10 ,,	+7°	+0·014° C. (0·0252° F.)	
11½ ,,	+9°	+0·018° C. (0·0324° F.)	
13 ,,	+13°	+0·026° C. (0·0468° F.)	
16 ,,	+15°	+0·03° C. (0·054° F.)	
18 ,,	+14°	+0·028° C. (0·0504° F.)	
20 ,,	+15·5°	+0·031° C. (0·0558° F.)	
24 ,,	+13·5°	+0·027° C. (0·0486° F.)	
26 ,,	+17°	+0·034° C. (0·0612° F.)	
29 ,,	+16°	+0·032° C. (0·0576° F.)	
30 ,,	+17°	+0·034° C. (0·0612° F.)	
31 ,,	+18°	+0·036° C. (0·0648° F.)	
31½ ,,	+19°	+0·038° C. (0·0684° F.)	
32 ,,	+20°	+0·04° C. (0·072° F.)	
33 ,,	+17°	+0·034° C. (0·0612° F.)	
34 ,,	+19°	+0·038° C. (0·0684° F.)	
35 ,,	+20°	+0·04° C. (0·072° F.)	Stopped reading.
37 ,,	+18°	+0·036° C. (0·0648° F.)	
39 ,,	+16°	+0·032° C. (0·0576° F.)	
39½ ,,	+14°	+0·028° C. (0·0504° F.)	

In this observation reading aloud of poetry caused a rise of temperature of 0·04° C. (0·072° F.), in thirty-two minutes.

MIDDLE REGION.
1st Experiment.
Examination of 1st—2nd district, 4th tier, left side.

At the end of—

0 minutes	0°	0°	Commenced reading aloud poetry.
2½ ,,	+3°	+0·006° C. (0·0108° F.)	

Temperature of the Head.

Time from commencement of work.	Rise or fall of temperature.		Mental condition.
	Deflections of galvanometer.	Thermometric values.	

At the end of—

4½ minutes	+5·5°	+0·011° C. (0·0198° F.)	
5½ ,,	+6·5°	+0·013° C. (0·0234° F.)	
7 ,,	+9°	+0·018° C. (0·0324° F.)	
8 ,,	+10°	+0·02° C. (0·036° F.)	
9 ,,	+11°	+0·022° C. (0·0396° F.)	
10 ,,	+11·5°	+0·023° C. (0·0414° F.)	
11 ,,	+14°	+0·028° C. (0·0504° F.)	
12 ,,	+15°	+0·03° C. (0·054° F.)	
14 ,,	+16°	+0·032° C. (0·0576° F.)	
14½ ,,	+ ,,	+ ,, ,,	
15 ,,	+13°	+0·026° C. (0·0468° F.)	} Attention failing.
15⅔ ,,	+14°	+0·028° C. (0·0504° F.)	
16 ,,	+15·5°	+0·031° C. (0·0558° F.)	Attention again aroused.
16½ ,,	+16°	+0·032° C. (0·0576° F.)	
17½ ,,	+16·5°	+0·033° C. (0·0594° F.)	
18½ ,,	+17°	+0·034° C. (0·0612° F.)	
19½ ,,	+19°	+0·038° C. (0·0684° F.)	
20 ,,	+ ,,	+ ,, ,,	
20½ ,,	+18·5°	+0·037° C. (0·0666° F.)	
22 ,,	+19·5°	+0·039° C. (0·0702° F.)	
24 ,,	+22·5°	+0·045° C. (0·081° F.)	
28 ,,	+23°	+0·046° C. (0·0828° F.)	

In this experiment the temperature of the space examined rose 0·046° C. (0·0828° F.), in twenty-eight minutes.

2nd Experiment.

Examination of 2nd district, 3rd tier, left side.

At the end of—

0 minutes	0°	0°	Reading of poetry to subject commenced.
2½ ,,	+7°	+0·014° C. (0·0252° F.)	
5 ,,	+15°	+0·03° C. (0·054° F.)	
6½ ,,	+16°	+0·032° C. (0·0576° F.)	
7 ,,	+15·5°	+0·031° C. (0·0558° F.)	
9 ,,	+18·5°	+0·037° C. (0·0666° F.)	
11 ,,	+19·5°	+0·039° C. (0·0702° F.)	
14 ,,	+20°	+0·04° C. (0·072° F.)	
17 ,,	+21·5°	+0·043° C. (0·0774° F.)	
20 ,,	+21°	+0·042° C. (0·0756° F.)	
22 ,,	+22°	+0·044° C. (0·0792° F.)	
23 ,,	+ ,,	+ ,, ,,	

General Effect of Emotional Activity. 181

Time from commence- ment of work.	Rise or fall of temperature.		Mental condition.
	Deflections of galvanometer.	Thermometric values.	

At the end of—
- 25 minutes . +24° . +0·048° C. (0·0864° F.)
- 28½ ,, . +21° . +0·042° C. (0·0756° F.) Attention failing.
- 30 ,, . +23° . +0·046° C. (0·0828° F.)

The listening to the reading caused, in this case, a rise of 0·048° C. (0·0864° F.), in twenty-five minutes.

POSTERIOR REGION.

1st *Experiment*.

Examination of 5th district, 3rd tier, left side.

At the end of—
- 0 minutes . 0° . 0° Commenced reading aloud
- 3 ,, . +4° . +0·008° C. (0·0144° F.) poetry.
- 5 ,, . +4·5° . +0·009° C. (0·0162° F.)
- 8 ,, . +6° . +0·012° C. (0·0216° F.)
- 9½ ,, . + ,, . + ,, ,,
- 12 ,, . +9° . +0·018° C. (0·0324° F.)
- 15 ,, . +10° . +0·02° C. (0·036° F.)
- 18 ,, . +10·5° +0·021° C. (0·0378° F.)
- 22 ,, . +11° . +0·022° C. (0·0396° F.)
- 25 ,, . + ,, . + ,, ,,
- 27 ,, . +12° . +0·024° C. (0·0432° F.)
- 29 ,, . +13° . +0·026° C. (0·0468° F.)
- 30 ,, . +15° . +0·03° C. (0·054° F.)
- 31 ,, . +15·5° +0·031° C. (0·0558° F.)
- 32 ,, . +16° . +0·032° C. (0·0576° F.)
- 33 ,, . + ,, . + ,, ,,
- 35 ,, . +15·5° +0·031° C. (0·0558° F.)

The temperature of the head rose, in the above experiment, 0·032° C. (0·0576° F.), in thirty-two minutes.

2nd *Experiment*.

Examination of 4th district, 2nd tier, left side.

At the end of—
- 0 minutes . 0° . 0° Reading of poetry to
- 4 ,, . +3° . +0·006° C. (0·0108° F.) subject commenced.
- 6 ,, . +4·5° . +0·009° C. (0·0162° F.)
- 8 ,, . +5° . +0·01° C. (0·018° F.)

Time from commencement of work.	Rise or fall of temperature.		Mental condition.
	Deflections of galvanometer.	Thermometric values.	
At the end of—			
10 minutes	+6°	+0·012° C. (0·0216° F.)	
12 ,,	+8°	+0·016° C. (0·0288° F.)	
15 ,,	+9°	+0·018° C. (0·0324° F.)	
18 ,,	+10°	+0·02° C. (0·036° F.)	
20 ,,	+9°	+0·018° C. (0·0324° F.)	
22 ,,	+11°	+0·022° C. (0·0396° F.)	
25 ,,	+10·5°	+0·021° C. (0·0378° F.)	
26 ,,	+11°	+0·022° C. (0·0396° F.)	
28 ,,	+12°	+0·024° C. (0·0432° F.)	
30 ,,	+13°	+0·026° C. (0·0468° F.)	
32 ,,	+13·5°	+0·027° C. (0·0486° F.)	
34 ,,	+14·5°	+0·029° C. (0·0522° F.)	
35 ,,	+ ,,	+ ,, ,,	
36 ,,	+12°	+0·024° C. (0·0432° F.)	} Attention failing.
37 ,,	+10·5°	+0·021° C. (0·0378° F.)	
39 ,,	+12°	+0·024° C. (0·0432° F.)	} Attention again aroused.
41 ,,	+13°	+0·026° C. (0·0468° F.)	

In this experiment the temperature rose 0·029° C. (0·0522° F.), in thirty-four minutes.

If we take the same spaces which were selected in the experiments on intellectual work (p. 141), the average rises of temperature for each region due to the reading of, or the listening to, poetry and prose of an emotional character, are as follows :

Anterior region . . . 0·0385° C. (0·0693° F.)
Middle ,, . . . 0·041° C. (0·0738° F.)
Posterior ,, . . . 0·036° C. (0·0648° F.)

Average for all three regions 0·0385° C. (0·0693° F.).

In emotional states, as in the case of intellectual exertion, rises of temperature considerably above the average may at times be seen; the rise sometimes exceeding 0·1° C. (0·18° F.).

CHAPTER VII.

COMPARATIVE EFFECT OF EMOTIONAL ACTIVITY ON THE TEMPERATURE OF DIFFERENT SPACES OF ONE AND THE SAME SIDE OF THE HEAD.—COMPARATIVE EFFECTS OF EMOTIONAL ACTIVITY WITH AND WITHOUT VERBAL EXPRESSION.

FOLLOWING the course pursued with intellectual work, we will, in the next place, compare directly certain spaces with each other. The spaces selected are those already chosen for a similar purpose with reference to intellectual activity (p. 146), experience having shown that the same spaces are affected relatively very nearly the same in both kinds of mental action. As in the former experiments, the 1st—2nd district, 4th tier, middle region, is compared in turn with each of the other spaces, this space showing, usually, the greatest rise of temperature in emotional activity, as well as in intellectual work. Each experiment lasted thirty minutes.

Comparison of anterior and middle regions.

1st set of experiments.

Mean of 12 observations; 7 on left side, 5 on right side.

Comparison of $\begin{cases} \text{1st—2nd district, 4th tier, middle region.} \\ \text{5th district, 3rd tier, anterior region.} \end{cases}$

Rise of temperature of 0·003° C. (0·0054° F.) in favour of middle region.

2nd set of experiments.

Mean of 12 observations; 8 on left side, 4 on right side.

Comparison of $\begin{cases} \text{1st—2nd district, 4th tier, middle region.} \\ \text{5th district, 1st—2nd tier, anterior region.} \end{cases}$

Rise of temperature of 0·012° C. (0·0216° F.) in favour of middle region.

3rd set of experiments.

Mean of 12 observations; 8 on left side, 4 on right side.

Comparison of $\begin{cases} \text{1st—2nd district, 4th tier, middle region.} \\ \text{3rd district, 3rd tier, anterior region.} \end{cases}$

Rise of temperature of 0·012° C. (0·0216° F.) in favour of middle region.

4th set of experiments.

Mean of 12 observations; 8 on left side, 4 on right side.

Comparison of $\begin{cases} \text{1st—2nd district, 4th tier, middle region.} \\ \text{3rd—4th district, 5th tier, anterior region.} \end{cases}$

Rise of temperature of 0·013° C. (0·0234° F.) in favour of middle region.

5th set of experiments.

Mean of 12 observations; 7 on left side, 5 on right side.

Comparison of $\begin{cases} \text{1st—2nd district, 4th tier, middle region.} \\ \text{2nd district, 2nd tier, anterior region.} \end{cases}$

Rise of temperature of 0·014° C. (0·0252° F.) in favour of middle region.

Comparison of 1st—2nd district, 4th tier, middle region, with other spaces of the same region.

1st set of experiments.

Mean of 12 observations; 8 on left side, 4 on right side.

Comparison of $\begin{cases} \text{1st—2nd district, 4th tier.} \\ \text{3rd district, 4th tier.} \end{cases}$

Rise of temperature of 0·003° C. (0·0054° F.) in favour of 1st—2nd district.

2nd set of experiments.

Mean of 12 observations; 7 on left side, 5 on right side.

Comparison of $\begin{cases} \text{1st—2nd district, 4th tier.} \\ \text{3rd district, 2nd tier.} \end{cases}$

Rise of temperature of 0·011° C. (0·0198° F.) in favour of 1st—2nd district, 4th tier.

3rd set of experiments.

Mean of 12 observations; 7 on left side, 5 on right side.

Comparison of $\begin{cases} \text{1st—2nd district, 4th tier.} \\ \text{3rd district, 5th tier.} \end{cases}$

Rise of temperature of 0·012° C. (0·0216° F.) in favour of 1st—2nd district, 4th tier.

4th set of experiments.

Mean of 12 observations; 7 on left side, 5 on right side.

Comparison of $\begin{cases} \text{1st—2nd district, 4th tier.} \\ \text{4th district, 7th tier.} \end{cases}$

Rise of temperature of 0·013° C. (0·0234° F.) in favour of 1st—2nd district, 4th tier.

Comparison of middle and posterior regions.

1st set of experiments.

Mean of 12 observations; 8 on left side, 4 on right side.

Comparison of $\begin{cases} \text{1st—2nd district, 4th tier, middle region.} \\ \text{5th district, 3rd tier, posterior region.} \end{cases}$

Rise of temperature of 0·018° C. (0·034° F.) in favour of middle region.

2nd set of experiments.

Mean of 12 observations; 8 on left side, 4 on right side.

Comparison of $\begin{cases} \text{1st—2nd district, 4th tier, middle region.} \\ \text{4th district, 2nd tier, posterior region.} \end{cases}$

Rise of temperature of 0·02° C. (0·036° F.) in favour of middle region.

3rd set of experiments.

Mean of 12 observations; 7 on left side, 5 on right side.

Comparison of $\begin{cases} \text{1st—2nd district, 4th tier, middle region.} \\ \text{2nd district, 4th tier, posterior region.} \end{cases}$

Rise of temperature of 0·022° C. (0·0396° F.) in favour of middle region.

Temperature of the Head.

4th set of experiments.

Mean of 12 observations; 8 on left side, 4 on right side.

Comparison of { 1st—2nd district, 4th tier, middle region.
2nd district, 6th tier, posterior region.

Rise of temperature of 0·022° C. (0·0396° F.) in favour of middle region.

It is shown by the above comparisons that here, as in the case of intellectual work, the 1st—2nd district, 4th tier, middle region, exhibits the highest rise of temperature of all the spaces compared. Calling the value of the absolute rise of temperature in this space 1000, we have the following comparative values for the rises of temperature in the other spaces:

Values of absolute rises of temperature.

Anterior region.

5th district, 3rd tier	. . . .	928.
3rd „ 3rd tier	. . . .	714.
5th „ 1st—2nd tier	. . .	714.
3rd—4th district, 5th tier	. . .	690.
2nd district, 2nd tier	. . . .	666.

Middle region.

1st—2nd district, 4th tier	. . .	1000.
3rd district, 4th tier	. . . .	928.
3rd „ 2nd „	. . . .	738.
3rd „ 5th „	. . . .	714.
4th „ 7th „	. . . .	690.

Posterior region.

5th district, 3rd tier	. . . .	571.
4th „ 2nd „	. . . .	524.
2nd „ 4th „	. . . .	476.
2nd „ 6th „	. . . .	476.

As in the case of intellectual work, the highest rises of temperature are found in the 1st—2nd, and 3rd districts, 4th tier, middle region, and the 5th district, 3rd tier, anterior region. Indeed, the tract of highest rise of temperature marked out for intellectual work holds good equally for emotional activity. It is, however, usually much more difficult, and often impossible, to detect inequality of rise of temperature, in emotional conditions, in the spaces showing the greatest thermal activity. This fact led the author to consider, for some time, that no such area of superior rise of temperature as had been found for intellectual work, existed in the case of emotional states of the mind. Under favorable circumstances it can, however, be shown, that the portion of the head specified exhibits, in emotional conditions of the kind under consideration, a greater rise of temperature than other parts, the degree of this superiority, as well as the frequency of its existence, being, however, less than in the case of intellectual exertion.

The following are examples of comparisons of spaces, where the inequalities of rise of temperature were well marked.

Comparison of anterior and middle regions.

1st Experiment.

Comparison of 1st—2nd district, 4th tier, middle region, with 5th district, 1st—2nd tier, anterior region; both spaces on left side.

$+$ signifies in favour of middle region.
$-$ signifies in favour of anterior region.

Time from commence-ment of work.	Rise of temperature.		Mental condition.
	Deflections of galvanometer.	Thermometric values.	
At the end of—			
0 minutes	$0°$	$0°$	Reading of poetry to subject commenced.
5 ,,	$+4°$	$+0·008°$ C. ($0·0144°$ F.)	
7 ,,	$+5°$	$+0·01°$ C. ($0·018°$ F.)	
10 ,,	$+6·5°$	$+0·013°$ C. ($0·0234°$ F.)	
12 ,,	$+7°$	$+0·014°$ C. ($0·0252°$ F.)	
13 ,,	$+$,,	$+$,, ,,	
16 ,,	$+6·5°$	$+0·013°$ C. ($0·0234°$ F.)	
18 ,,	$+7°$	$+0·014°$ C. ($0·0252°$ F.)	
20 ,,	$+4°$	$+0·008°$ C. ($0·0144°$ F.)	
22 ,,	$+$,,	$+$,, ,,	Interest flagging.
23 ,,	$+3°$	$+0·006°$ C. ($0·0108°$ F.)	

Temperature of the Head.

	Rise of temperature.		
Time from commencement of work.	Deflections of galvanometer.	Thermometric values.	Mental condition.
At the end of—			
25 minutes	+5°	+0·01° C. (0·018° F.)	Interest returning.
27 ,,	+6·5°	+0·013° C. (0·0234° F.)	
29 ,,	+7°	+0·014° C. (0·0252° F.)	
32 ,,	+6·5°	+0·013° C. (0·0234° F.)	
			Stopped reading.
34 ,,	+5·5°	+0·011° C. (0·0198° F.)	
36 ,,	+4·5°	+0·009° C. (0·0162° F.)	
37 ,,	+3°	+0·006° C. (0·0108° F.)	
39 ,,	−1°	−0·002° C. (0·036° F)	

Highest rise, in favour of middle region, 0·014°C. (0·0252° F.).

2nd Experiment.

Comparison of 1st—2nd district, 4th tier, middle region, with 3rd district, 3rd tier, anterior region; both spaces on left side.

+ signifies in favour of middle region.
− signifies in favour of anterior region.

At the end of—			
0 minutes	0°	0°	Commenced reading aloud poetry.
3½ ,,	+3°	+0·006° C. (0·0108° F.)	
4 ,,	+4°	+0·008° C. (0·0144° F.)	
5⅔ ,,	+3°	+0·006° C. (0·0108° F.)	
6⅔ ,,	+5°	+0·01° C. (0·018° F.)	
7½ ,,	+5·5°	+0·011° C. (0·0198° F.)	
8 ,,	+7°	+0·014° C. (0·0252° F.)	
8½ ,,	+8°	+0·016° C. (0·0288° F.)	
10 ,,	+3°	+0·006° C. (0·0108° F.)	
12 ,,	+ ,,	+ ,, ,,	
13½ ,,	+2°	+0·004° C. (0·0072° F.)	
15 ,,	+5°	+0·01° C. (0·018° F.)	Interest flagging.
16 ,,	+6°	+0·012° C. (0·0216° F.)	
19 ,,	+1°	+0·002° C. (0·0036° F.)	
21 ,,	+2°	+0·004° C. (0·0072° F.)	
22 ,,	+ ,,	+ ,, ,,	
22½ ,,	+4°	+0·008° C. (0·0144° F.)	
			Stopped reading.
24 ,,	+1°	+0·002° C. (0·0036° F.)	
27 ,,	+3°	+0·006° C. (0·0108° F.)	
28 ,,	+ ,,	+ ,, ,,	
29 ,,	−1·5°	−0·003° C. (0·0054 F.)	
30 ,,	−2°	−0·004° C. (0·0072° F.)	

Highest rise, in favour of middle region, 0·016°C. (0·0288° F.).

3rd Experiment.

Comparison of 1st—2nd district, 4th tier, middle region, with 5th district, 3rd tier, anterior region; both spaces on left side.

+ signifies in favour of middle region.
− signifies in favour of anterior region.

Time from commencement of work.	Deflections of galvanometer.	Rise of temperature. Thermometric values.	Mental condition.
At the end of—			
0 minutes	0°	0°	Reading of poetry to subject commenced.
3 ,,	+2°	+0·004° C. (0·0072° F.)	
6 ,,	+3°	+0·006° C. (0·0108° F.)	
8½ ,,	+4°	+0·008° C. (0·0144° F.)	
10 ,,	+3°	+0·006° C. (0·0108° F.)	
12 ,,	+1°	+0·002° C. (0·0036° F.)	
14 ,,	−2°	−0·004° C. (0·0072° F.)	
15 ,,	−4·5°	−0·009° C. (0·0162° F.)	
16 ,,	−4°	−0·008° C. (0·0144° F.)	
17 ,,	− ,,	− ,, ,,	
18 ,,	−3°	−0·006° C. (0·0108° F.)	
20 ,,	0°	0°	
21 ,,	+2°	+0·004° C. (0·0072° F.)	
23 ,,	+4°	+0·008° C. (0·0144° F.)	
25 ,,	+6°	+0·012° C. (0·0216° F.)	
28 ,,	+ ,,	+ ,, ,,	
30 ,,	+5·5°	+0·011° C. (0·0198° F.)	
32 ,,	+6°	+0·012° C. (0·0216° F.)	
34 ,,	+ ,,	+ ,, ,,	Stopped reading.
36 ,,	+5°	+0·01° C. (0·018° F.)	
38 ,,	+3°	+0·006° C. (0·0108° F.)	
40 ,,	+2°	+0·004° C. (0·0072° F.)	
41 ,,	0°	0°	

In the above experiment, during the first eight and a half minutes, the rise of temperature in the middle region predominated; but, setting out from this time, the deflection of the galvanometer fell back to zero, and then passed over to the side in favour of the anterior region. This movement of the needle was not owing to a fall of temperature in the middle region (as was proved by direct testing), but to an increase in the activity of thermal production in the anterior region. At the end of the fifteenth minute from the commencement of the experiment, the rise of temperature in favour of the anterior region was 0·009° C. (0·0162° F.). After this time, the middle region began anew to gain upon the anterior region, and, at the end of the twenty-first minute, the balance of temperature was again in favour of the former region. At the close of the reading the superiority of rise in favour of the middle region was very marked, for the particular two spaces examined, being 0·012° C. (0·0216° F.).

Comparison of 1st—2nd district, 4th tier, middle region, with other spaces of the same region.

1st Experiment.

Comparison of 1st—2nd district, 4th tier, with 3rd district, 5th tier; both spaces on left side.

  + signifies in favour of 1st—2nd district, 4th tier.
  − signifies in favour of 3rd district, 5th tier.

Time from commencement of work.	Rise of temperature.		Mental condition.
	Deflections of galvanometer.	Thermometric values.	
At the end of—			
0 minutes	0°	0°	Reading of poetry to subject commenced.
4 ,,	+3°	+0·006° C. (0·0108° F.)	
6 ,,	+4°	+0·008° C. (0·0144° F.)	
7 ,,	+4·5°	+0·009° C. (0·0162° F.)	
8½ ,,	+4°	+0·008° C. (0·0144° F.)	
10 ,,	+4·5°	+0·009° C. (0·0162° F.)	
12 ,,	+5°	+0·01° C. (0·018° F.)	
14 ,,	+5·5°	+0·011° C. (0·0198° F.)	
			Stopped reading.
17 ,,	+3°	+0·006° C. (0·0108° F.)	
18 ,,	+2°	+0·004° C. (0·0072° F.)	
			Commenced again reading.
21 ,,	+4°	+0·008° C. (0·0144° F.)	
22 ,,	+4·5°	+0·009° C. (0·0162° F.)	
26 ,,	+5°	+0·01° C. (0·018° F.)	
29 ,,	+6°	+0·012° C. (0·0216° F.)	
31 ,,	+6·5°	+0·013° C. (0·0234° F.)	
33 ,,	+ ,,	+ ,,	,,

Highest rise, in favour of 1st—2nd district, 5th tier, 0·013°C. (0·0234° F.).

Comparative Effect of Emotional Activity. 191

2nd Experiment.

Comparison of 1st—2nd district, 4th tier, with 3rd district, 4th tier; both spaces on left side.

+ signifies in favour of 1st—2nd district.
— signifies in favour of 3rd district.

Time from commencement of work.	Deflections of galvanometer.	Thermometric values.	Mental condition.
At the end of—			
0 minutes	0°	0°	Reading of poetry to subject commenced.
3 ,,	+1°	+0·002° C. (0·0036° F.)	
5 ,,	−1°	−0·002° C. (0·0036° F.)	
7 ,,	−3°	−0·006° C. (0·0108° F.)	
9 ,,	+1°	+0·002° C. (0·0036° F.)	
11 ,,	+3°	+0·006° C. (0·0108° F.)	
13 ,,	+4°	+0·008° C. (0·0144° F.)	
16 ,,	+3°	+0·006° C. (0·0108° F.)	
18 ,,	+4°	+0·008° C. (0·0144° F.)	
21 ,,	+ ,,	+ ,,	,,
24 ,,	+ ,,	+ ,,	,,

In this experiment the superiority of rise of temperature was, at the start, for a moment in favour of the 1st—2nd district; but from the end of the third to the end of the seventh minutes, the 3rd district showed the higher rate of rise. After the seventh minute, the 1st—2nd district again began to rise the more rapidly, and finally, at the end of the thirteenth minute, its superiority of rise over the other space was 0·008° C. (0·0144° F.).

3rd Experiment.

Comparison of 1st—2nd district, 4th tier, with 3rd district, 2nd tier; both spaces on left side.

+ signifies in favour of 1st—2nd district, 4th tier.
— signifies in favour of 3rd district, 2nd tier.

At the end of—			
0 minutes	0°	0°	Reading aloud of poetry commenced.
3½ ,,	+3°	+0·006° C. (0·0108° F.)	
5 ,,	+3·5°	+0·007° C. (0·0126° F.)	
7 ,,	+4°	+0·008° C (0·0144° F.)	
9 ,,	+5°	+0·01° C. (0·018° F.)	
12 ,,	+6·5°	+0·013° C. (0·0234° F.)	
15 ,,	+7°	+0·014° C. (0·0252° F.)	

Temperature of the Head.

		Rise of temperature.		
Time from commencement of work.	Deflections of galvanometer.	Thermometric values.		Mental condition.

At the end of—

16 minutes	.	+6·5°.	+0·013° C. (0·0234° F.)	
18 ,,	.	+ ,,	. + ,, ,,	
21 ,,	.	+6°.	+0·012° C. (0·0216° F.)	
23 ,,	.	+6·5°.	+0·013° C. (0·0234° F.)	
				Stopped reading.
25 ,,	.	+5°.	+0·01° C. (0·018° F.)	
30 ,,	.	+3°.	+0·006° C. (0·0108° F.)	
35 ,,	.	0°.	0°.	
36 ,,	.	−2°.	−0·004° C. (0·0072° F.)	
37 ,,	.	−3°.	−0·006° C. (0·0108° F.)	

In this experiment the superiority of rise of temperature was in favour of the 1st—2nd district, 4th tier, the highest point attained being 0·014° C. (0·0252° F.). The apparent rise in favour of the 3rd district, 2nd tier, after the cessation of the reading, was, in reality, a more rapid fall of temperature in the 1st—2nd district, 4th tier, than that occurring simultaneously in the 3rd district, 2nd tier.

Comparison of middle and posterior regions.

1st Experiment.

Comparison of 1st—2nd district, 4th tier, middle region, with 2nd district, 4th tier, posterior region; both spaces on left side.

+ signifies in favour of middle region.

At the end of—

0 minutes	.	0°.	0°	Commenced reading poetry
3 ,,	.	+2°.	+0·004° C. (0·0072° F.)	to subject.
6 ,,	.	+4°.	+0·008° C. (0·0144° F.)	
8 ,,	.	+5°.	+0·01° C. (0·018° F.)	
11½ ,,	.	+6°.	+0·012° C. (0·0216° F.)	
14 ,,	.	+8·5°.	+0·017° C. (0·0306° F.)	
16 ,,	.	+9°.	+0·018° C. (0·0324° F.)	
19 ,,	.	+ ,,	. + ,,	
22 ,,	.	+9·5°.	+0·019° C. (0·034° F.)	
24 ,,	.	+10°.	+0·02° C. (0·036° F.)	
25 ,,	.	+11°.	+0·022° C. (0·0396° F.)	
27 ,,	.	+10°.	+0·02° C. (0·036° F.)	
29 ,,	.	+10·5°	+0·021° C. (0·0378° F.)	

Comparative Effect of Emotional Activity. 193

Time from commence-ment of work.	Rise of temperature.		Mental condition.
	Deflections of galvanometer.	Thermometric values.	

At the end of—
31 minutes . +10° . +0·02° C. (0·036° F.)
               Stopped reading.
39 ,, . +5° . +0·01° C. (0·018° F.)
42 ,, . +3° . +0·006° C. (0·0108° F.)

Highest rise, in favour of middle region, 0·022° C. (0·0396° F.).

2nd Experiment.

Comparison of 1st—2nd district, 4th tier, middle region, with 5th district, 3rd tier, posterior region; both spaces on left side.

+ signifies in favour of middle region.

At the end of—
0 minutes . 0° . 0°   Reading of poetry aloud
3½ ,, . +1° . +0·002° C. (0·0036° F.)   commenced.
4 ,, . +2° . +0·004° C. (0·0072° F.)
6 ,, . +5° . +0·01° C. (0·018° F.)
9 ,, . +5·5°. +0·011° C. (0·0198° F.)
12 ,, . + ,, . + ,,   ,,
15 ,, . +6° . +0·012° C. (0·0216° F.)
17 ,, . + ,, . + ,,
20 ,, . +6·5°. +0·013° C. (0·0234° F.)
23 ,, . +7° . +0·014° C. (0·0252° F.)
25 ,, . + ,, . + ,,   ,,
27 ,, . +8° . +0·016° C. (0·0288° F.)
28 ,, . +9° . +0·018° C. (0·0324° F.)
31 ,, . +8·5°. +0·017° C. (0·0306° F.)
35 ,, . +9° . +0·018° C. (0·0324° F.)
37 ,, . +8·5°. +0·017° C. (0·0306° F.)
               Stopped reading.
39 ,, . +7·5°. +0·015° C. (0·027° F.)
43 ,, . +4·5°. +0·009° C. (0·0162° F.)
45 ,, . +3·5°. +0·007° C. (0·0126° F.)

Highest rise, in favour of middle region, 0·018° C. (0·0324° F.).

Reversals of the usual order of superiority of rise of temperature in the different spaces are still more frequent in emotional activity than in intellectual work.

It has been stated (p. 123) that recitation to one's self produces a greater effect on the temperature of the head than recitation aloud. The only difficulty that lies in the way of making

Temperature of the Head.

comparative experiments on this point, is the keeping of the attention and interest from flagging without the aid of the sound of the voice. Practice, however, may overcome this difficulty, and then the result will be as stated. The difference in the two cases is slight, but in taking the averages of a number of observations it is sufficiently evident; it may vary from 0·002° C. (0·0036° F.) to 0·008° C. (0·0144° F.). The comparative effects of the two kinds of recitation may sometimes be obtained in one and the same experiment, as in the following example:

Examination of 1st—2nd district, 4th tier, middle region, left side.

+ signifies rise of temperature above the starting point.

Time from commence- ment of work.	Rise of temperature.		Mental condition.
	Deflections of galvanometer.	Thermometric values.	
At the end of—			
0 minutes	0°	0°	Commenced recitation of poetry aloud.
2 ,,	+3°	+0·006° C. (0·0108° F.)	
5 ,,	+7°	+0·014° C. (0·0252° F.)	
8 ,,	+9·5°	+0·019° C. (0·0342° F.)	
11 ,,	+11°	+0·022° C. (0·0396° F.)	
13 ,,	+12°	+0·024° C. (0·0432° F.)	
18 ,,	+16°	+0·032° C. (0·0576° F.)	
22 ,,	+16·5°	+0·033° C. (0·0594° F.)	
23 ,,	+17°	+0·034° C. (0·0612° F.)	
25 ,,	+18°	+0·036° C. (0·0648° F.)	
27 ,,	+ ,,	+ ,,	,,
28 ,,	+17·5°	+0·035° C. (0·063° F.)	
30 ,,	+19°	+0·038° C. (0·0684° F.)	
32 ,,	+18°	+0·036° C. (0·0648° F.)	
34 ,,	+ ,,	+ ,,	,,
			Stopped recitation aloud and commenced repeating to one's self.
35 ,,	+18·5°	+0·037° C. (0·0666° F.)	
39 ,,	+19·5°	+0·039° C. (0·0702° F.)	
42 ,,	+20·5°	+0·041° C. (0·0738° F.)	
44 ,,	+21·5°	+0·043° C. (0·0774° F.)	
45½ ,,	+22°	+0·044° C. (0·0792° F.)	
			Commenced recitation aloud again.
47 ,,	+21°	+0·042° C. (0·0756° F.)	
49½ ,,	+20·5°	+0·041° C. (0·0738° F.)	
52 ,,	+19·5°	+0·039° C. (0·0702° F.)	

Comparative Effect of Emotional Activity.

Time from commencement of work.	Rise of temperature.		Mental condition.
	Deflections of galvanometer.	Thermometric values.	
At the end of—			
54 minutes	+18·5°	+0·037° C. (0·0666° F.)	
59 „	+20·5°	+0·041° C. (0·0738° F.)	Stopped recitation aloud and commenced repeating to one's self.
64 „	+21·5°	+0·043° C. (0·0774° F.)	
66 „	+ „	+ „ „	
67 „	+ „	+ „ „	
70 „	+20·5°	+0·041° C. (0·0738° F.)	Commenced recitation aloud again.
75 „	+19°	+0·039° C. (0·0684° F.)	
77 „	+ „	+ „ „	
82 „	+17°	+0·034° C. (0·0612° F.)	Stopped reciting.
87 „	+13°	+0·026° C. (0·0468° F.)	
90 „	+10°	+0·02° C. (0·036° F.)	

In the first part of this experiment recitation aloud caused a maximum rise of 0·038° C. (0·0684° F.). On substituting recitation to one's self for recitation aloud, the temperature rose to 0·044° C. (0·0792° F.). A return to recitation aloud caused the temperature to fall back to 0·037° C. (0·0666° F.). On again commencing to repeat to one's self, a renewal of the rise of temperature took place, the highest point now attained being 0·043° C. (0·0774° F.). Another return to recitation aloud diminished the last maximum to 0·038° C. (0·0684° F.).*

* This greater effect of recitation to one's self than of recitation aloud would appear to be in accordance with the laws of the correlation and conservation of force. In internal recitation an additional portion of energy, which in recitation aloud was converted into nervous and muscular force, now appears as heat. The author believes that this explanation was suggested to him by Professor George F. Barker, of Yale College, U.S., in 1867. It is to be found in a pamphlet by Professor Barker, entitled 'The Correlation of Vital Physical Forces,—published at New Haven, Connecticut, 1871.

CHAPTER VIII.

COMPARATIVE EFFECT OF EMOTIONAL ACTIVITY ON THE TEMPERATURE OF SYMMETRICALLY SITUATED SPACES OF THE TWO SIDES OF THE HEAD.—EXAMPLES OF THE EFFECTS OF ANGER, VEXATION, AND MIRTH, ON THE TEMPERATURE OF THE HEAD.

WE pass next to the comparative effect of emotional activity on the temperature of the two sides of the head. The spaces compared are the same as those examined under similar circumstances in intellectual work, with one omission,—the 3rd—4th district, 5th tier, anterior region.

ANTERIOR REGION.

1st Series of Experiments.
Comparison of 2nd district, 2nd tier, on the two sides.
12 observations.

Times of occurrence of superiority of rise of temperature on a side, and of equality of rise of temperature on the two sides.	Mean difference of rise of temperature observed at the end of 30 minutes' recitation aloud.
In favour of—	In favour of—
Left side . 8	Left side . 0·007° C. (0·0126° F.)
Right ,, . 2	Right ,, . 0·006° C. (0·0108° F.)
Equality . 2	

2nd Series of Experiments.
Comparison of 3rd district, 3rd tier, on the two sides.
12 observations.

In favour of—	In favour of—
Left side . 8	Left side . 0·007° C. (0·0126° F.)
Right ,, . 3	Right ,, . 0·005° C. (0·009° F.)
Equality . 1	

Comparative Effect of Emotional Activity. 197

3rd Series of Experiments.
Comparison of 5th district, 3rd tier, on the two sides.
12 observations.

Times of occurrence of superiority of rise of temperature on a side, and of equality of rise of temperature on the two sides.

Mean difference of rise of temperature observed at the end of 30 minutes' recitation aloud.

In favour of—
Left side . 9
Right ,, . 2
Equality . 1

In favour of—
Left side . 0·008° C. (0·0144° F.)
Right ,, . 0·006° C. (0·0108° F.)

4th Series of Experiments.
Comparison of 5th district, 1st—2nd tier, on the two sides.
12 observations.

In favour of—
Left side . 8
Right ,, . 2
Equality . 2

In favour of—
Left side . 0·008° C. (0·0144° F.)
Right ,, . 0·006° C. (0·0108° F.)

MIDDLE REGION.

1st Series of Experiments.
Comparison of 1st—2nd district, 4th tier, on the two sides.
12 observations.

In favour of—
Left side . 7
Right ,, . 2
Equality . 3

In favour of—
Left side . 0·007° C. (0·0126° F.)
Right ,, . 0·006° C. (0·0108° F.)

2nd Series of Experiments.
Comparison of 3rd district, 2nd tier, on the two sides.
12 observations.

In favour of—
Left side . 7
Right ,, . 3
Equality . 2

In favour of—
Left side . 0·005° C. (0·009° F.)
Right ,, . 0·005° C. (0·009° F.)

3rd Series of Experiments.
Comparison of 3rd district, 5th tier, on the two sides.
12 observations.

Times of occurrence of superiority of rise of temperature on a side, and of equality of rise of temperature on the two sides.

Mean difference of rise of temperature observed at the end of 30 minutes' recitation aloud.

In favour of—
Left side . 8
Right „ . 2
Equality . 2

In favour of—
Left side . 0·0055° C. (0·0099° F.)
Right „ . 0·004° C. (0·0072° F.)

4th Series of Experiments.
Comparison of 4th district, 6th tier, on the two sides.
12 observations.

In favour of—
Left side . 7
Right „ . 3
Equality . 2

In favour of—
Left side . 0·005° C. (0·009° F.)
Right „ . 0·0045° C. (0·0081° F.)

POSTERIOR REGION.

1st Series of Experiments.
Comparison of 2nd district, 4th tier, on the two sides.
12 observations.

In favour of—
Left side . 6
Right „ . 3
Equality . 3

In favour of—
Left side . 0·004° C. (0·0072° F.)
Right „ . 0·003° C. (0·0054° F.)

2nd Series of Experiments.
Comparison of 2nd district, 6th tier, on the two sides.
12 observations.

In favour of—
Left side . 7
Right „ . 3
Equality . 2

In favour of—
Left side . 0·004° C. (0·0072° F.)
Right „ . 0·004° C. (0·0072° F.)

3rd Series of Experiments.
Comparison of 4th district, 2nd tier, on the two sides.
12 observations.

Times of occurrence of superiority of rise of temperature on a side, and of equality of rise of temperature on the two sides.

Mean difference of rise of temperature observed at the end of 30 minutes' recitation aloud.

In favour of—
Left side . 6
Right ,, . 3
Equality . 3

In favour of—
Left side . 0·005° C. (0·009° F.)
Right ,, . 0·005° C. (0·009° F.)

4th Series of Experiments.
Comparison of 5th district, 3rd tier, on the two sides.
12 observations.

In favour of—
Left side . 6
Right ,, . 3
Equality . 3

In favour of—
Left side . 0·0055° C. (0·0099° F.)
Right ,, . 0·005° C. (0·009° F.)

The above experiments show that, in the greater number of cases, in all three regions, the rise of temperature during emotional activity is higher on the left side than on the right side. In a certain number of instances, however, as in the case of intellectual work, the right side shows the higher temperature, or both sides rise equally. Here, as in the experiments on intellectual work, the anterior region surpasses the other two regions, both as regards the number of instances of greater rise of temperature in favour of the left side, and also as regards the degree of difference of rise observed on both sides. The order of the middle and posterior regions is also the same in these respects as in intellectual work.

The following figures give the percentages of results in favour of the left and right sides, and of equality of the two sides, respectively, and also the mean differences of rise of temperature in favour of each side in each region.

Anterior Region.

	Left side.	Right side.	Equality.
Average percentages of times of occurrence of superiority of rise of temperature on a side, and of equality of rise of temperature on the two sides	68·75	18·75	12·5

Average difference of rise } 0·0075° C. . 0·00575° C.
of temperature on a side . } (0·0135° F.) (0·01035° F.)

Middle Region.

	Left side.	Right side.	Equality.
Average percentages of times of occurrence of superiority of rise of temperature on a side, and of equality of rise of temperature on the two sides	60·4167	20·8333	18·75

Average difference of } 0·00562° C. . 0·00487° C.
rise of temperature . } (0·0116° F.) (0·00877° F.)

Posterior Region.

	Left side.	Right side.	Equality.
Average percentages of times of occurrence of superiority of rise of temperature on a side, and of equality of rise of temperature on the two sides	52·0833	25·000	22·9167

Average difference of rise } 0·00462° C. 0·00425° C.
of temperature on a side . } (0·00831° F.) (0·00765° F.)

Of the total 144 observations, 87, or 60·416 per cent., are in favour of the left side; 31, or 21·528 per cent., are in favour of the right side; and 26, or 18·056 per cent., are in favour of equality of the two sides.

Comparing the averages of the results obtained in emotional

activity with those of the same class obtained in intellectual work (p. 172), we see,—first, that in emotional activity the average difference of rise of temperature in favour of both sides of the head, in every region, is greater than in intellectual work; second, that the percentages of times of occurrence of superiority of rise of temperature on the left side are greater, in all three regions, in intellectual work than in emotional activity; third, that the higher percentages of results in favour of superiority of rise of temperature on the right side, and also of equality of rise of temperature on the two sides, are found, in all three regions, in emotional activity.

In the kind of mental activity which we are now considering, alterations of the position of the balance of superiority of temperature on the two sides of the head, are much more frequent and more marked than in intellectual work. As in the case of the latter, these alterations usually tend to increase the extent of the tract of superior temperature on the left side. There is no regularity in these changes of the position of higher temperature, sometimes one set of spaces being affected and sometimes another. They are most marked, as in the case of intellectual work, in the anterior region, but are also not unfrequent in the middle and posterior regions.

In all the experiments of the class now under consideration, the effect of alterations of the circulation, both general and local, and of the muscular exertion involved in recitation aloud, must be taken into account. The last mentioned influence may indeed be left out of consideration, since, as has been before stated, mere mechanical recitation without emotion produces no effect, while recitation to one's self, as before remarked, is more potent in raising the temperature than recitation aloud. With regard to the effect of alterations of the general circulation, it is found that in comparing points over large vessels with the head, the latter shows the greater rise of temperature, or rather the *only* rise, as the other points are usually unaffected. Points situated over the femoral artery in the groin, over the brachial artery, and over the carotid artery (in the latter instance during recitation to one's self only), have been examined for this purpose.

With reference to local vascular alterations, due to vaso-motor influence, there can be no doubt that they play a more or less important part in the rise of temperature in question;

but how prominent this part is it is impossible at present to determine.*

The effect of anger, in a moderate degree, on the normal distribution of temperature, has been examined in four instances, all in the same individual. The examinations were made when the first intensity of passion had subsided, and the mental condition might be said to be that of indignation. There was no alteration of colour in the face, at the time of the observations, and the circulation and respiration had quieted down. In one of the cases nothing definite could be determined;—there existed simply that confusion of distribution of temperature which seems, in a greater or lesser degree, to generally precede a change in the relative temperatures of the two sides, and which, in some persons, is brought about by even slight mental excitement. In the other three instances, however, the effects were well defined, and persisted long enough to allow of a pretty careful inspection. Of course, in these cases, the only method of examination applicable was to test the relative temperatures of different spaces during the fit of indignation, and to compare the results with those which previous examinations had obtained in the quiescent mental condition. The head was also examined after the mind had resumed its former state of tranquility. Fortunately, in two instances the head had been examined but a short time before the access of the fit of anger, in a number of spaces. The changes observed were as follows :

ANTERIOR REGION.

1st Case.—The whole of the left side of higher temperature than the right side, with the exception of the 2nd and 3rd districts, 5th tier.

2nd Case.—The whole of the left side of higher temperature than the right side, with the exception of the 2nd and 3rd districts, 4th and 5th tiers.

3rd Case.—The whole of the left side of higher temperature than the right side, with the exception of the 3rd and 4th districts 4th tier, and the 3rd district, 5th and 6th tiers.

* The author believes that all the higher degrees of rise of temperature seen at the surface of the head during both intellectual and emotional activity, are in part owing to vascular disturbance. The rise of temperature of nearly $0.5°$ C. ($0.9°$ F.), observed by M. Broca after ten minutes' reading aloud, would appear to be partly due to the above cause.

Middle Region.

1st Case.—The whole of the 2nd and 3rd tiers, usually of higher temperature on the right side, now of higher temperature on the left side.

2nd Case.—The 1st, 2nd, and 3rd districts of the 2nd and 3rd tiers, previously of higher temperature on the right side, now of higher temperature on the left side.

3rd Case.—The 1st and 2nd districts of the 2nd and 3rd tiers, and the 4th and 5th districts of the 3rd tier, previously of higher temperature on the right side, now of higher temperature on the left side.

Posterior Region.

1st Case.—The 3rd, 4th, and 5th districts of the 2nd and 3rd tiers, usually of higher temperature on the right side, now of higher temperature on the left side.

2nd Case.—Uncertain.

3rd Case.—The 4th and 5th districts of the 2nd and 3rd tiers, previously of higher temperature on the right side, now of higher temperature on the left side.

Quantitative results could not be obtained owing to the haste necessary in the observations in order to compare as many spaces as possible before the mental condition changed.

But, in a number of instances, fits of anger, irritability and annoyance have occurred in the very midst of experiments both on the absolute and also on the comparative rise of temperature in different spaces from increased mental activity,—thus affording opportunities of judging of the effects of the above mental conditions both on the absolute and relative temperatures of certain parts of the head. The following are examples of the effects of the conditions of the mind in question on the absolute temperatures of certain spaces:

1st Experiment.

Examination of 3rd district, 3rd tier, anterior region, left side.

+ signifies rise of temperature above starting point.

Time from commencement of work.	Rise of temperature. Deflections of galvanometer.	Thermometric values.	Mental condition.
At the end of—			
0 minutes	0°	0°	Commenced mathematical work.
15 ,,	+12°	+0·024° C. (0·0432° F.)	
17 ,,	+10·5°	+0·021° C. (0·0378° F.)	Work interrupted, and subject greatly annoyed.
19 ,,	+12°	+0·024° C. (0·0432° F.)	
22 ,,	+15°	+0·03° C. (0·054° F.)	
25 ,,	+16°	+0·032° C. (0·0576° F.)	
			Annoyance disappearing.
26 ,,	+15°	+0·03° C. (0·054° F.)	
29 ,,	+13°	+0·026° C. (0·0468° F.)	
31 ,,	+12°	+0·024° C. (0·0432° F.)	
			Annoyance completely gone, and work resumed.
33 ,,	+ ,,	+ ,, ,,	
37 ,,	+11·5°	+0·023° C. (0·0414° F.)	

In this experiment annoyance caused a rise of temperature of 0·011° C. (0·0298° F.), (from 0·021° C. to 0·032° C.) in the eight minutes of its continuance.

2nd Experiment.

Examination of 3rd district, 2nd tier, middle region, right side.

+ signifies rise of temperature above starting point.

At the end of—			
10 minutes	+3°	+0·006° C. (0·0108° F.)	Idle, after a few minutes' reading aloud. Angry.
11 ,,	+ ,,	+ ,, ,,	
12 ,,	+6°	+0·012° C. (0·0216° F.)	
13 ,,	+6·5°	+0·013° C. (0·0238° F.)	
14 ,,	+7°	+0·014° C. (0·0252° F.)	
15 ,,	+6°	+0·012° C. (0·0216° F.)	Anger passing away.
16 ,,	+4°	+0·008° C. (0·0144° F.)	
18 ,,	+2°	+0·004° C. (0·0072° F.)	Anger entirely gone.
19 ,,	+1°	+0·002° C. (0·0036° F.)	

We have here, as the result of anger of short duration, a rise of temperature of 0·008° C. (0·0144° F.).

Effects of Anger, Vexation, &c.

3rd Experiment.

Examination of 3rd district, 4th tier, middle region, left side.

+ signifies rise of temperature above starting point.

Time from commencement of observation.	Rise of temperature.		Mental condition.
	Deflections of galvanometer.	Thermometric values.	
At the end of—			
6 minutes	+5°	+0·01° C. (0·018° F.)	Had been engaged in conversation of interest since the commencement of the experiment, and was beginning to be irritated.
8 ,,	+9°	+0·018° C. (0·0324° F.)	
10 ,,	+10°	+0·02° C. (0·036° F.)	
13 ,,	+14°	+0·028° C. (0·0504° F.)	
14 ,,	+15°	+0·03° C. (0·054° F.)	
17 ,,	+15·5°	+0·031° C. (0·0558° F.)	Much vexed.
20 ,,	+16°	+0·032° C. (0·0576° F.)	
22 ,,	+ ,,	+ ,, ,,	
23 ,,	+17°	+0·034° C. (0·0612° F.)	Conversation ended, but vexation continuing.
26 ,,	+ ,,	+ ,, ,,	
27 ,,	+ ,,	+ ,, ,,	
30 ,,	+16°	+0·032° C. (0·0576° F.)	
34 ,,	+16·5°	+0·033° C. (0·0594° F.)	Vexation passed away.
39 ,,	+ ,,	+ ,, ,,	
43 ,,	+15·5°	+0·031° C. (0·0558° F.)	
50 ,,	+12°	+0·024° C. (0·0432° F.)	
55 ,,	+8°	+0·016° C. (0·0288° F.)	

In this experiment the temperature which had been raised in the first place by conversation, was further and decidedly increased by vexation, this increase being 0·024° C. (0·0432° F.). Moreover, in this case, after the complete disappearance of the mental disturbance, the temperature still remained above its first level at the commencement of the observation, falling only very slowly.

We will next consider the cases in which anger or vexation has occurred during the comparison of two spaces.

206   *Temperature of the Head.*

1st Experiment.

Comparison of 3rd district, 4th tier, middle region, with 4th district, 2nd tier, posterior region; both spaces on left side.

+ signifies in favour of middle region.

	Rise of temperature.	

Time from commencement of work.	Deflections of galvanometer.	Thermometric values.	Mental condition.
At the end of—			
0 minutes	$0°$	$0°$	Commenced reading aloud.
30 ,,	$+8°$	$+0·016°$ C. ($0·0288°$ F.)	At this point the subject
32 ,,	$+9°$	$+0·018°$ C. ($0·0324°$ F.)	became angry and the
34 ,,	$+12°$	$+0·024°$ C. ($0·0432°$ F.)	reading stopped.
36 ,,	$+11°$	$+0·022°$ C. ($0·0396°$ F.)	
38 ,,	$+$,,	$+$,,	,,
40 ,,	$+$,,	$+$,,	,, Anger passed away.
42 ,,	$+9°$	$+0·018°$ C. ($0·0324°$ F.)	
43 ,,	$+7°$	$+0·014°$ C. ($0·0252°$ F.)	

In this case the occurrence of anger increased the already existing superiority of rise in the middle region due to reading, by $0·008°$ C. ($0·0144°$ F.). It will be noticed that the temperature, after the thirty-fourth minute, appeared to fall slightly. It was found, however, that no fall had taken place in the middle region, but that the posterior region was rising in temperature as well as the middle region. Both spaces were, therefore, affected by the mental state,—the space of the middle region, however, in the greater degree.

2nd Experiment.

Comparison of 1st—2nd district, 4th tier, middle region, with 5th district, 3rd tier, anterior region; both spaces on left side.

+ signifies in favour of middle region.
− signifies in favour of anterior region.

At the end of—			
0 minutes	$0°$	$0°$	Subject became much vexed.
3 ,,	$+1°$	$+0·002°$ C. ($0·0036°$ F.)	
4 ,,	$+2°$	$+0·004°$ C. ($0·0072°$ F.)	
5 ,,	$−1°$	$−0·002°$ C. ($0·0036°$ F.)	
7 ,,	$0°$	$0°$	
8 ,,	$+1°$	$+0·002°$ C. ($0·0036°$ F.)	Vexation entirely gone,
9 ,,	$0°$	$0°$	and subject commencing
12 ,,	$+3°$	$+0·006°$ C. ($0·0108°$ F.)	to read.
14 ,,	$+4°$	$+0·008°$ C. ($0·0144°$ F.)	

In this experiment, after the appearance of vexation, the temperature oscillated on both sides of zero. This irregularity was proved to be due to inequalities in the rates of rise of the two spaces from one moment to another; both spaces were rising in temperature, but the rapidity of rise was alternately greater in each, the maximum rise being, however, in favour of the middle region.

3rd Experiment.

Comparison of 1st—2nd district, 4th tier, middle region, with 3rd district, 3rd tier, anterior region, both spaces on left side.

+ signifies in favour of middle region.

Time from commencement of work.	Rise of temperature.		Mental condition.
	Deflections of galvanometer.	Thermometric values.	
At the end of—			
0 minutes	0^c	$0°$	Commenced reading aloud.
25 ,,	$+7°$	$+0·014°$ C. ($0·0252°$ F.)	
27 ,,	$+9^c$	$+0·018°$ C. ($0·0324°$ F.)	Stopped reading; much vexed.
29 ,,	$+10°$	$+0·02°$ C. ($0·036°$ F.)	
30 ,,	$+$,,	$+$,,	,,
32 ,,	$+9°$	$+0·018°$ C. ($0·0324°$ F.)	
35 ,,	$+8°$	$+0·016°$ C. ($0·0288°$ F.)	Vexation disappearing.
37 ,,	$+7·5°$	$+0·015°$ C. ($0·027°$ F.)	
40 ,,	$+3°$	$+0·006°$ C. ($0·0108°$ F.)	Vexation entirely gone.

Here vexation increased the superiority of rise of temperature in the middle region already existing, by $0·006°$ C. ($0·0108°$ F.).

From the above experiments, and from others of a similar character, it would seem not unlikely that the comparative effect, on the temperature of different spaces, of anger and vexation follows a similar rule to that usually applicable to intellectual work and the particular kind of emotional activity which we have principally studied.*

* These possible effects of anger and vexation show the danger of trusting implicitly to results obtained on unknown subjects. For a considerable length of time the author was not aware that one of his most valuable subjects was liable to fits of ill-temper, which sometimes persisted for many hours. After having learned this fact, it was easy to explain discrepancies in results previously obtained, and—to a certain extent—to foretell the results of observations made

On two occasions vexation has occurred during experiments on the relative rise of temperature of the two sides of the head. In one of the experiments (examination of 1st—2nd district, 4th tier, middle region), the rise of temperature which ensued was greater on the left side by 0·003° C. (0·0054° F.). In the other case (examination of 3rd district, 3rd tier, anterior region) the rise was equal on the two sides.

The effect of moderate mirth on the temperature of two spaces (3rd district, 3rd tier, anterior region, and 1st—2nd district, 4th tier, middle region; both spaces on left side) has been witnessed on four occasions; it has consisted in a slight rise of temperature,—0·003° C. (0·0054° F.) to 0·005° C. (0·009° F.).; but with regard to the relative degree of this rise in different spaces, nothing has as yet been satisfactorily ascertained.

In closing this, the last of the chief divisions of our investigations, a few words must be said with regard to the application of the results given to the question involved in the second of the two ultimate objects of the researches as set forth on the first page of the Introduction.

We have seen that all parts of the surface of the head—even when taken in such small subdivisions as those which we have adopted—show an increase of temperature during all kinds of mental work; but that some parts are usually more active in this respect than others; although, here again, the different parts seem to be able to supply each other's places,—the commonly less active part not unfrequently superseding its ordinary superior. This much established, we are at once brought back to the old question, discussed at the close of Part II, of how far the temperatures of points of the outer surface represent the temperatures of points of cerebral tissue lying directly beneath them; and the answer is, as before, that there is no certainty that the relative

when the condition of the mind in question existed. One of two things is generally found in such cases—namely, either, that no effect at all is produced by mental work—probably because the subject cannot attend properly—or that the temperature *falls*—the degree of thermal activity attending the mental work being less than that accompanying the previous emotional condition. Of the risks of error in examining the relative temperatures of the different spaces under such circumstances, the examples given on pages 202, 203, are sufficient evidence.

Conclusion.

temperatures of small subdivisions of the outer surface represent with exactitude the relative temperatures of the underlying tracts of brain surface, but that it is highly probable that in the case of larger areas—such as, for example, the group of spaces specified as showing the highest rises of temperature—a definite relation exists between the two classes of values : hence, the inference is, that—taking the rise of temperature as the best available measure of functional activity—the relative elevations of temperature of areas of a certain size of the outer surface, during mental exercise, represent with considerable correctness the relative degrees of functional activity of the corresponding underlying portions of brain surface.

APPENDIX.

DIAGRAM 7.

POINTS of the cerebral convolutions underlying some of the most important of the subdivisions of the surface of the head adopted in the present work. Figure 3 (from Bitot, 'Essai de Topographie Cérébrale').

S.F.P.	—	Fronto-parietal suture.
S.R.	—	Fissure of Rolando.
S.i.P.	—	Interparietal fissure.
S.S.	—	Fissure of Sylvius.
S.P.	—	Parallel fissure.
S.F.i.	—	Internal frontal fissure.
S.F.e.	—	External ,, ,,
S.Pr.	—	Precentral fissure.
4.F.	—	Fourth frontal convolution [anterior ascending parietal].
1.F.	—	First frontal convolution.
2.F.	—	Second ,, ,,
3.F.	—	Third ,, ,,
C.P.A.	—	Ascending parietal convolution [posterior ascending parietal].
L.P.s.	—	Superior parietal lobule.
L.P.i.	—	Inferior ,, ,,
1.T., 2.T., 3.T.	—	First, second, and third temporal convolutions.

A.	—	5th district, 1st tier, anterior region.			
B.	—	5th	,,	2nd tier	,, ,,
D.	—	3rd	,,	3rd tier	,, ,,
G.	—	4th	,,	,,	,, ,,
H.	—	5th	,,	,,	,, ,,
I.	—	4th	,,	4th tier	,, ,,